Complete Guide on GENERAL AGRICULTURE

The Authors

Mr. Deepak Kumar Mahanta, presently pursuing his Ph.D from Dr. RPCAU, PUSA, Bihar in the Department of Agricultural Entomology. He has completed his graduation from ANGRAU, Bapatla, A.P. through ICAR. He got All India 5th rank and awarded with JRF by ICAR. He has completed his M.Sc. (Ag) in the Department of Entomology, Indian Agricultural Research Institute, New Delhi. He has qualified ICAR-NET in the year 2021 also qualified SRF-2021 (AIR-7). He has secured Best thesis award in SPPS national conference, ICAR-IARI, New Delhi.

Ms. J. Komal, presently pursuing her Ph.D. from Navsari Agricultural University in the field of Agricultural Entomology. She had graduated with distinction from Agricultural College, Aswaraopeta (PJTSAU) in 2019. She has got University Silver Medal for highest O.G.P.A in Plant Protection. She had secured all India 3rd rank in Entomology and awarded with JRF by ICAR and completed her Master's in Entomology from ICAR-Indian Agricultural Research Institute. She has secured prestigious IARI gold medal. Later, secured all India 8th rank in SRF Crop sciences exam by ICAR. Qualified ICAR-NET. She has secured Best thesis award in SPPS national conference, ICAR-IARI, New Delhi.

Mr. Ajesh B. R., did his B.Sc. (Agriculture) from Agricultural College, Bapatla, ANGRAU, MSc (Plant Pathology) from MPKV, Rahuri and currently pursuing his Ph.D. (Plant Pathology) in Tamil Nadu Agricultural University, Coimbatore. He has published research-based articles in international journal and presented paper in national symposium. He believes that disseminating knowledge is the human duty and sharing it about so that all can benefit.

Dr. Md. Abbas Ahmad is presently working as Assistant Professor-cum-Scientist in the Department of Entomology, P.G. College of Agriculture, Dr. Rajendra Prasad Central Agricultural University, Pusa (Bihar). He obtained his Doctoral degree from G.B. Pant University of Agriculture and Technology, Pantnagar. He has more than 15 years of experience in Teaching and Research. He has several National and International publications. He is now continuing his research on All India Co-ordinated Rice Improvement Project.

Complete Guide on GENERAL AGRICULTURE

For all the examinations conducted by ASRB, ARS, SAUs, JRF, SRF, UGC-NET, UPSC, PCS, ICMR, CSIR-NET, NAARM, DBT, NDRI, IARI, IVRI and Various Universities

Deepak Kumar Mahanta

J. Komal

Ajesh B.R.

Md. Abbas Ahmad

2024

Astral Cracker

A Division of

Astral International Pvt. Ltd.

New Delhi – 110 002

ISBN: 978-93-5461-977-9 (Paperback)

Published by : **Astral Cracker**
A Division of
Astral International Pvt. Ltd.
– ISO 9001:2015 Certified Company –
4736/23, Ansari Road, Darya Ganj,
New Delhi-110 002
Ph. 011-43549197, 8130496929
E-mail: info@astralint.com
Website: www.astralint.com

Digitally Printed at : **Replika Press Pvt. Ltd.**

Acknowledgement

I, Deepak Kumar Mahanta, feel actuated from within to offer my humblest and sincerest thanks to the almighty God, the merciful the beneficent, who bestowed me the ability to pursue high ideals in life. To be able to step strong and smooth in this way, I have also been supported and supervised by many people to whom I would like to express my earnest gratitude. These two people I mention deserve much credit for guiding me through life and making me the person I am today. My father Mr. Kapilendra Mahanta and mother Smt. Urmila Mahanta who has always been there through thick and thin, backing supporting, adoring, defending me and cannot imagine my existence without their presence. Thank you for your pain, patience, sunshine, memories, kindness and everything that you two bestowed towards me. I would like to give special thanks to my beloved mamu Mr. Bhubaneshwar Mahanta and Mami Lopamudra Mahanta for their parental care and supporting in each and every step of life. I will be always thankful to Ms J. Komal for her ever-supporting hand. I also thank my beloved sister Mrs. Debasmita Mahanta for her love and care. I owe thanks to Saiteja, Naidu, Dinesh, Babu, Ajesh, Ankit, Abhisek, Somananda Tanmaya, Ipsita for their needful help to write this book.

Deepak Kumar Mahanta

Preface

Agriculture being the backbone of our civilization attracts much attention towards a new revolution of its kind and it is the responsibility of upcoming generation to bring agriculture to a new high. The pandemic covid-19 made us realize the statement of Jawaharlal Nehru, '*Everything can wait but not agriculture.*' Realizing the importance and potency of agriculture, knowing its all the core subjects importance we have written this book with utmost care.

This book comprises several core subjects of agriculture like Agronomy, Soil science, Entomology, Plant pathology, Genetics, Plant physiology and social sciences. This book will be useful for the candidates preparing for the competitive examinations like ICAR-JRF, ICAR-SRF, UPSC, ICFRE, CSB, AFO (Mains) and several other state examinations.

I humbly take an opportunity to thank my family, teachers, friends and seniors for the moral support they have given. I would extend my gratitude to all the authors of the book for their timeless efforts and cooperation. It would be my pleasure to express thankfulness to the publishers of Astral publication for their smooth and tireless efforts to publish the book in a faster pace without compromising the quality.

Deepak Kumar Mahanta

J. Komal

Ajesh B. R

Dr. Md. Abbas Ahmad

Contents

SOIL SCIENCE

PLANT BREEDING AND GENETICS

AGRONOMY

Chapter 1

ICAR, National and International Institutions, History of Ancient and Modern Agriculture, Evolution of Agriculture, Contribution of Agricultural Experts, Important Days Related to Agriculture, Revolutions in Agriculture, different Schemes of Agriculture

1. **By 2022, How many National research centres and project directorates are present in the system of ICAR?**

   (a) 14, 13 (b) 15,13

   (c) 13,15 (d) 14,12

2. **By 2022, How many KVKs are present in system of ICAR? Maximum nos of KVKs are present in which ATARI?**

   (a) 698, Pune (b) 724, Kanpur

   (c) 712, Jabalpur (d) 730, Nagpur

3. **Who is called Father of economic ecology?**

   (a) Dr. S.K Vasal (b) Dr. V Kurien

   (c) Dr. M.S. Swaminathan (d) Dr. N.E. Borlaug

4. **Which of the following honour award is not received by M.S Swaminathan?**

   (a) Padma shree (b) Padma Bhushan

   (c) Padma vibhushan (d) Bharat Ratna

5. Who recently received 7th Dr. M. S Swaminathan award?

(a) Dr. V. Praveen Rao
(b) Dr. Trilochan Mohapatra
(c) Dr. T.K Behera
(d) Dr. S.K Singh

6. The Crop present in ICAR- LOGO is?

(a) Wheat
(b) Rice
(c) Banana
(d) Onion

7. The motto of ICAR in English is?

(a) Agrisearch with a human touch
(b) Agridevelopment is our development
(c) Agriculture is our culture
(d) None of the above

8. Who is Director general and President of ICAR 2022?

(a) Dr. Trilochan Mohapatra and Narendra Singh tomar
(b) Dr. Trilochan Mohapatra and Radha Mohan singh
(c) Narendra Singh tomar and Dr. Trilochan Mohapatra
(d) Dr. Trilochan Mohapatra and Narendra Modi

9. First ever cultivated crop in the history of Agriculture is?

(a) Rice
(b) Maize
(c) Potato
(d) Wheat and Barley

10. IMD was established at Pune in 1875 for meteorological survey in India. What is the full form of IMD?

(a) Indian Meteorological Department
(b) India Meteorological Department
(c) International meteorological Department
(d) Department of meteorological survey of India

11. The term Ever green revolution is coined by?

(a) William Gaud
(b) M.S. Swaminathan
(c) Amrita Patel
(d) Sanjay Rajaram

12. Who have received Nobel peace prize in Agriculture?

(a) N.E Borlaug
(b) M.S. Swaminathan
(c) K.L Chadda
(d) R.S Paroda

13. Which prize is considered as Nobel prize in Agriculture?
(a) Millennium prize
(b) Wolf Prize
(c) World food prize
(d) Choudhary charan shingh award

14. In 2021, recently who has received World food prize and in which field?
(a) Shakuntala Haraksingh Thilsted (aquatic food nutrition)
(b) Shakuntala Haraksingh Thilsted (Soil science)
(c) Ratan lal (Soil Science)
(d) Ratan lal (women nutrition)

15. PM Modi's Tricolour revolution or second green revolution includes
(a) White revolution (milk)
(b) Saffron (energy)
(c) Blue revolution (clean water and fishes)
(d) All of the above

16. Operation green extended to
(a) Fruits
(b) Vegetables
(c) Both
(d) Green manure crops

17. Golden rice contains
(a) Beta carotene
(b) Beta xanthene
(c) Anthocyanin
(d) Alfa carotene

18. Which year is considered as International year of millets by FAO
(a) 2021
(b) 2022
(c) 2023
(d) 2020

19. Kishan day is celebrated on
(a) 23rd December
(b) 5th December
(c) 15th March
(d) 20 January

20. New name of Rashtriya krishi Vikas Yojana (RKVY) is
(a) RKVY-TOTAL
(b) RKVY-ATARI
(c) RKVY-SVANIDHI
(d) RKVY-RAFTAAR

21. Quality protein maize is rich in which amino acids?
(a) Glycine and tryptophan
(b) Lysine and tryptophan
(c) Lysine and leucine
(d) Methionine and tryptophan

22. Father of super rice and super wheat are?

(a) Khush and Nagarajan
(b) Nagarajan and Khush
(c) Vasal and Chadda
(d) Paroda and Sanjay Rajaram

23. India's first pollinators park is opened at

(a) Trissur, Kerala
(b) Bhubaneshwar, Odisha
(c) Ganganagar, Rajasthan
(d) Nainital, UK

24. India's food grain production in crop year 2020-2021 is

(a) 300 million tonnes
(b) 309 million tonnes
(c) 295 million tonnes
(d) 320 million tonnes

25. In which state 1st grain ATM and milk ATM started in INDIA respectively?

(a) Haryana, Gujarat
(b) Gujarat, Haryana
(c) Maharashtra, Uttar Pradesh
(d) Uttar Pradesh, Maharashtra

26. Water soluble vitamins are

(a) A, D, E, K
(b) B and C
(c) A, B, C, D
(d) D and E

27. Prabhani revolution is related to

(a) Okra production
(b) Tomato production
(c) Fertilizer production
(d) None of the above

28. International centre for maize and wheat improvement is situated at

(a) Philippines
(b) Mexico
(c) Taiwan
(d) Nigeria

29. National bureau of agriculturally important micro-organism is situated in which state?

(a) Maharashtra
(b) Jammu and Kashmir
(c) Nagaland
(d) Uttar Pradesh

30. Which of the following is incorrect?

(a) Vasal- maize
(b) Chadda- fruit
(c) Khush-rice
(d) Sanjay Rajaram- rice

31. 1st AICRP and 1st NRC started for?

(a) Maize and rice
(b) Maize and wheat
(c) Maize and groundnut
(d) Maize and Barley

32. Fungicide is derived from which word?

(a) Greek (b) Latin

(c) French (d) Hindi

33. Who has won the miss universe title 2021 from India

(a) Shushmita Sen (b) Alia Bhatt

(c) Rashmika Mandhana (d) Harnaaz Sandhu

34. The headquarter of WMO is situated at?

(a) Geneva (b) Rome

(c) Kanbera (d) New Delhi

35. Which plant parts are useful for Isabgol?

(a) Husk (b) Seed

(c) Root (d) Both (a) and (b)

36. Which of the following is not the recipient of world food prize?

(a) N.E Borloug (b) M.S. Swaminathan

(c) S.K. Vashal (d) V. Kurien

37. Which transgenic plant covered highest area in the world?

(a) Maize (b) Cotton

(c) Canola (d) Soyabean

38. Who has awarded wolf Prize for extra ordinary contribution to plant breeding and genetics?

(a) S.K Vashal (b) G.S Khush

(c) N.E Borlaug (d) M.S Swaminathan

39. NABARD is established in the year

(a) 1982 (b) 1972

(c) 1992 (d) 1962

40. Which of the international institute is present in India?

(a) ICARDA (b) CRIDA

(c) ICRISAT (d) ILRAD

41. Degnala or alkali disease is caused due to mineral

(a) Se deficiency (b) Co deficiency

(c) Co excesses (d) None of the above

42. 2,4-D is used as fruit setting hormone at

(a) <20ppm (b) >20ppm

(c) <50ppm (d) None of the above

43. Sproutstop is trade name of?

(a) Mallic hydrazide (b) CCC

(c) Ethylene (d) Abscisic acid

44. NIAM is situated at?

(a) Raipur (b) Jaipur

(c) Cuttack (d) Nagpur

45. AVRDC is situated at?

(a) Rome, Italy (b) Sri Lanka

(c) Taiwan (d) Nigeria

46. Which football legend's statue has been unveiled in Panaji, Goa?

(a) Sunil Chhetri (b) Christiano Ronaldo

(c) Messi (d) Baichung bhutia

47. Financial Stability Report (FSR)' is the flagship report released by which institution?

(a) NITI Aayog (b) World bank

(c) RBI (d) Asian development bank

48. The GST Council has decided to defer the increase in GST Rate on which product?

(a) textiles (b) Automobile

(c) Mobile phones (d) Cotton

49. Which is India's first indigenously developed Receptor Binding Domain (RBD) protein sub-unit vaccine for COVID-19?

(a) CORBEVAXTM (b) NOVAVAX

(c) DIVIVAX (d) SEROVAXTM

50. As per APEDA, which is the top exported commodity in India in 2020-21?

(a) Basmati rice (b) Non-basmati rice

(c) Milk (d) Cotton

ANSWER KEY

1	2	3	4	5	6	7	8	9	10	11	12	13	14	15
(a)	(b)	(c)	(d)	(a)	(a)	(a)	(a)	(d)	(b)	(b)	(a)	(c)	(a)	(d)
16	17	18	19	20	21	22	23	24	25	26	27	28	29	30
(c)	(a)	(c)	(a)	(d)	(b)	(a)	(d)	(b)	(a)	(b)	(a)	(b)	(d)	(d)
31	32	33	34	35	36	37	38	39	40	41	42	43	44	45
(c)	(b)	(d)	(a)	(d)	(a)	(d)	(b)	(a)	(c)	(a)	(a)	(a)	(b)	(c)
46	47	48	49	50										
(b)	(c)	(a)	(a)	(b)										

Chapter 2

Different Types of Farming, Agronomy Introduction, Concept of different Terms like LAI, LER, Cropping Intensity, Cropping System, Harvesting Index, Agroclimatic and Agroecological Zones, Droughts, Agroforestry, Soil Conservation, Climate, Weather, Instruments and Uses, Monsoon, Soil Moisture, Critical Stages, Irrigation, Soil Moisture Constants, WUE

1. Power house of Nutrients/Nutri-Cereals/Climate resilient crops are?

(a) Cereals (b) Pulses

(c) Millets (d) Grasses

2. India's first ever brand logo for cotton and India's premier cotton would be known as?

(a) Supriya cotton (b) Kasturi cotton

(c) Samba cotton (d) Suruchi cotton

3. World's First conversation area for sea cucumber created at?

(a) Bhubaneshwar (b) Gujarat

(c) Lakshadweep (d) Kerala

4. _______ becomes the first state under Pradhan Mantri Fasal Bhima Yojana (PMFBY) to cover crop loss done by wild animals?

(a) Sikkim (b) Mizoram

(c) Nagaland (d) Tripura

5. **Online registration portal developed by National Bee Board under National Beekeeping and Honey Mission (NBHM) to achieve a traceability source of Honey and checking source and the quality and source of adulteration of honey**
   (a) Madhu seva (b) Madhurkranti
   (c) Sahad dan (d) Mahu grow

6. **The prediction of monsoon season based on analysis of weather data of premonsoon period and managing crops is?**
   (a) Eco farming (b) Jhum farming
   (c) Precision farming (d) Response farming

7. **The target specific use of inputs like seeds, fertilizer, pesticides for crop production according to crop requirements is known as**
   (a) Precision farming (b) Response farming
   (c) Terra farming (d) Shifting farming

8. **To develop the mars according to the Earth environment is called**
   (a) Response farming (b) Ideotype
   (c) Precision farming (d) Terra farming

9. **The model plant which is expected to perform or behave in a predictable manner with in defined environment**
   (a) Jhum (b) Biological plant
   (c) Ideotype (d) Type specimen

10. **Father of zero tillage is**
   (a) G.B Triplets (b) K. Rahman
   (c) K.l William (d) Amir khan

11. **A method aiming at reducing tillage to the minimum necessary for ensuring a good seedbed, rapid generation, satisfactory crop stand**
   (a) Zero tillage (b) Minimum tillage
   (c) Till planting (d) Least tillage

12. **Which of the following is not a greenhouse gas?**
   (a) Carbon dioxide (b) Methane
   (c) Nitrous oxide (d) Oxygen

13. **Cereals are deficient in amino acids?**
   (a) Lysine and threonine (b) Methionine and arginine
   (c) Lysine and arginine (d) Methionine and threonine

14. VAM is essential for

(a) K solubilization (b) P solubilization

(c) S solubilization (d) N solubilization

15. Monsoon responsible for rain in Tamil Nadu and Kerala is

(a) S-E monsoon (b) S-w monsoon

(c) N-E monsoon (d) N-W monsoon

16. Biofertilizer used in wheat and sorghum respectively

(a) Azatobactor and Azospirilum (b) Glomus and saprolegnia

(c) Azospirilum and azotobacter (d) Rhizobium and glomus

17. Carbon content and C:N Ratio for organic matter is?

(a) 48, 12:1 (b) 58, 10:1

(c) 58, 12:1 (d) 48, 10:1

18. State having highest area under summer maize?

(a) Bihar (b) Jharkhand

(c) Odisha (d) West Bengal

19. Evapotranspiration is measured by

(a) Tensiometer (b) Lysimeter

(c) Transpiometer (d) Barometer

20. The greenhouse gas released from rice fields is

(a) Nitrous oxide (b) Carbon dioxide

(c) Methane (d) CFC

21. Pusa giant is a variety of

(a) Berseem (b) Napier grass

(c) Papaya (d) All of the above

22. Aminophilic plants are

(a) Rice (b) Sugarcane

(c) Both (d) Tobacco

23. Rancidity of sunflower is due to

(a) Oxidation (b) Reduction

(c) Oxido-reduction (d) Hydrolization

24. Crop which has highest K^+ uptake is

(a) Tobbacco (b) Cotton

(c) Onion (d) Potato

25. Highest award presented to an agricultural scientist in the country

(a) Rafi Ahmed Kidwai Award

(b) Choudhary Charan Singh Award

(c) Hari Om Ashram Trust Award

(d) Jawaharlal Nehru Award

26. Father of soil microbiology?

(a) Traug (b) Leibig

(c) Winogradsky (d) Dokuchev

27. The word "weeds" and zero tillage is given by

(a) Triplett (b) Arthur young

(c) Watson (d) None of the above

28. The ratio between total leaf area to the ground area is known as

(a) LAI (b) LER

(c) CRI (d) CI

29. The ratio of yield of inter crop to the yield of sole crop is known as

(a) LAI (b) LER

(c) Cropping intensity (d) Crop rotation intensity

30. The ratio between the total cropping area to the net cultivated area is known as?

(a) Cropping intensity (b) LAI

(c) LER (d) Crop rotation intensity

31. The stubbles of sorghum takes long time to decompose due to wide C:N Ratio and temporary immobilization of nitrogen is known as

(a) Sorghum solidness (b) Sorghum sickness

(c) Sorghum rigidness (d) None of the above

32. Annual rainfall more than 1150mm is classified under

(a) Rainfed farming (b) Dryland farming

(c) Dry farming (d) None of the above

33. No of agroclimatic and agroecological zones present in India is?

(a) 20, 08 (b) 20,15

(c) 15,20 (d) 30,20

34. The example of stomatal closing type anti traspirants are

(a) Atrazine, CO_2 (b) Kaolin

(c) Mobileaf (d) Silicone

35. Reflecting type anti traspirants example

(a) Atrazine (b) Kaoline

(c) Mobileaf (d) Cycocel

36. Zing terracing is adopted when the slope is

(a) Less than 2 per cent (b) 3-10 per cent

(c) More than15 per cent (d) 16-33 per cent

37. Agronomic and mechanical measures are taken when slopes are respectively

(a) <2 per cent,>2 per cent (b) >2 per cent, <2 per cent

(c) 12 per cent, 15 per cent (d) 15 per cent, 12 per cent

38. Bench terracing is done when the slope is ranging from

(a) 3-10 per cent (b) 16-33 per cent

(c) 2-10 per cent (d) Less than 2 per cent

39. All the weather phenomenon like clouds, rain, fog, most, dew occur in

(a) Troposphere (b) Stratosphere

(c) Mesosphere (d) Thermosphere

40. Ozone layer (Ozonosphere) is present in

(a) Troposphere (b) Stratosphere

(c) Mesosphere (d) Thermosphere

41. Which gas is present in highest amount in atmospheric air?

(a) Neon (b) Xenon

(c) Argon (d) Krypton

42. The unit of solar constant is

(a) Watt/m^2 (b) Percentages

(c) Cal/cm^2/min (d) Both (a) and (c)

43. Lines of connecting points of equal amount of rainfall?

(a) Isobaths (b) Isobar

(c) Isohytes (d) Isohel

44. Specific gravity of soil is measured by

(a) Peizometer
(b) Pyranometer
(c) Crescograph
(d) Pycnometer

45. CRI is critical stage for

(a) Rice
(b) Wheat
(c) Gram
(d) Groundnut

46. 'Plan Bee', which was seen in the news recently, is associated with which Union Ministry?

(a) Ministry of Railways
(b) Ministry of Environment, Forest and Climate Change
(c) Ministry of Agriculture and Farmers Welfare
(d) Ministry of MSME

47. When is the 'Pravasi Bharatiya Diwas' observed in India?

(a) Januay1
(b) January 9
(c) January 14
(d) January 23

48. Which is the first city in India to have a water metro project?

(a) Kolkata
(b) Kochi
(c) Chennai
(d) Mumbai

49. As per a recent study, 'Huntsman spiders' present in which country eat tree frogs?

(a) Madagascar
(b) Singapore
(c) India
(d) Brazil

50. Which is India's first large state to record 100 per cent first-dose vaccinations?

(a) Telangana
(b) Odisha
(c) Punjab
(d) Uttar Pradesh

ANSWER KEY

1	2	3	4	5	6	7	8	9	10	11	12	13	14	15
(c)	(b)	(c)	(a)	(b)	(d)	(a)	(d)	(c)	(a)	(b)	(d)	(a)	(b)	(c)
16	17	18	19	20	21	22	23	24	25	26	27	28	29	30
(a)	(b)	(a)	(b)	(c)	(d)	(c)	(a)	(d)	(a)	(c)	(a)	(a)	(b)	(a)
31	32	33	34	35	36	37	38	39	40	41	42	43	44	45
(b)	(a)	(c)	(a)	(b)	(b)	(a)	(b)	(a)	(b)	(c)	(d)	(c)	(d)	(b)
46	47	48	49	50										
(a)	(b)	(b)	(a)	(a)										

Chapter 3

Crop Production of Paddy, Wheat, Barley, Maize, Bajra, Jowar, Sugarcane, Chickpea, Pigeon Pea

1. **India's First Kisan special Parcel Train or Kisan Rail transported fruits and vegetables between**
   (a) Gujarat to West Bengal
   (b) Maharashtra to Bihar
   (c) Punjab to Maharashtra
   (d) Karnataka to Tamil Nadu

2. **India's first honey testing laboratory set up at-**
   (a) Pune
   (b) Anand
   (c) Chandigarh
   (d) Bhubaneshwar

3. **The first country in the world to adopt a ban on deforestation, a ground-breaking move that could have a major global impact is**
   (a) India
   (b) Norway
   (c) America
   (d) Jamaica

4. **The first Indian state to declare state mangrove tree (*Sonneratia alba*) as symbol of conservation is**
   (a) Maharashtra
   (b) Odisha
   (c) Assam
   (d) West Bengal

5. **The First state in the country to implement "Land Leasing Policy"**
   (a) Uttarakhand
   (b) Jammu and Kashmir
   (c) Punjab
   (d) Odisha

6. Nipping is done in

(a) Gram
(b) Cotton
(c) Both
(d) Banana

7. Which is known as vegetable meat

(a) Cowpea
(b) Soyabean
(c) Mushroom
(d) Paneer

8. Highest and lowest water use efficiency is present respectively for

(a) Finger millet and bajra
(b) Pearl millet and maize
(c) Finger millet and rice
(d) Rice and finger millet

9. Downward movement of water through saturated and nearly saturated soil in response to gravity is known as?

(a) Seepage
(b) Percolation
(c) Infiltration
(d) Leaching

10. Season (sowing time) of Aman rice is

(a) April-may
(b) June-July
(c) November-December
(d) Jan-February

11. Which rice contains more amount of protein

(a) White rice
(b) Brown rice
(c) Polished rice
(d) All contain same

12. The circular burning patches is typical damage symptom of

(a) Zn deficiency
(b) BPH insect damage
(c) Nitrogen deficiency
(d) Blast disease

13. The West African rice is

(a) *O. sativa*
(b) *O. Nivara*
(c) *O. Glaberrima*
(d) None of the above

14. Test weight of rice, basmati rice, wheat and bajra are respectively? (in grams)

(a) 21,25, 5, 40
(b) 40,20,21,5
(c) 25,21,40,5
(d) 5,40,20,21

15. The variety developed by the cross between Peta x Dee gee woo gen is?

(a) TN 1
(b) JAYA
(c) IR 8
(d) IR 20

16. Which variety of rice is known as Miracle rice in India

(a) TN 1 (b) IR 8

(c) Jagannath (d) Jaya

17. Khaira disease of rice is caused due to

(a) Zn deficiency (b) Zn excesses

(c) Fe toxicity (d) Fe deficiency

18. Browning of rice is caused due to

(a) Iron toxicity (b) Iron deficiency

(c) Zinc deficiency (d) Zinc toxicity

19. The cross ploughing of the standing crop for weed control and proper aeration is known as

(a) Puddling (b) Ploughing

(c) Beushening (d) None of the above

20. Milling of rice cause which disease

(a) Night blindness (b) Scurvy

(c) Beri beri (d) Rickettsia

21. In Dapog method seedlings are ready for transplanting after how many days of sowing

(a) 11-14 days (b) 20-25 days

(c) 15-20 days (d) 7-9 days

22. Gene responsible for dwarfness in rice and wheat is respectively

(a) Norin 10 and Norin 8

(b) Norin 10 and dee-gee-woo-gen

(c) Dee-gee-woo-gen and Norin 10

(d) Norin 8 and TKM 6

23. Emmer wheat is?

(a) *T. aestivum* (b) *T. diccocum*

(c) *T. monococcum* (d) None of the above

24. Which of the following is diploid species of wheat?

(a) *T. aestivum* (b) *T. durum*

(c) *T. diccocum* (d) *T. aeglloploid*

25. Which of the following is triple gene dwarf variety?

(a) Sonalika (b) Choti lerma

(c) Kalyan sona (d) None of the above

26. Triticale is the cross between

(a) Wheat x Barley (b) Wheat x rye

(c) Rye x wheat (d) Wheat x bajra

27. Which variety of wheat is resistant to all 3 types of rust diseases

(a) Sonalika (b) Lal badshah

(c) Choti lerma (d) Pusa lerma

28. The huskless variety of barly is

(a) Rajkiran (b) Ratna

(c) Dolma (d) Kailash

29. Malt is produced from grains of

(a) Rice (b) Barely

(c) Bajra (d) Wheat

30. White bud in maize is caused by

(a) Zinc toxicity (b) Zinc deficiency

(c) Phosphorus toxicity (d) Phosphorus deficiency

31. Double cross techniques of production is introduced by

(a) East and Shull (b) Jones

(c) Tripplet (d) Watson

32. Highest amount of minerals are present in which cereals?

(a) Rice (b) Wheat

(c) Bajra (d) Sorghum

33. Among cereals and millets which crop is highest drought tolerant?

(a) Rice (b) Wheat

(c) Bajra (d) Jowar

34. Sorghum crop should be used as fodder

(a) At knee stage (b) After flowering stage

(c) Before flowering stage (d) Can be used in any stage

35. The part of the Sugar cane used (best) for seed purposes is

(a) Top 1/3 to 1/2 portion (b) Bottom 1/3 to 1/2 portion

(c) Any part is best (d) Full top portion

36. Sugarcane is considered as matured if the brix ratio is

(a) 6-8 (b) 10-12

(c) 14-16 (d) 16-18

37. Most critical stage for irrigation in sugarcane is

(a) Germination (b) Formitive

(c) Grand growth (d) Maturity

38. Wilt resistant variety of chickpea is

(a) Avarodhi (b) Gaurav

(c) Aparna (d) Radhey

39. Extra early maturing variety of redgram is

(a) Upasana (b) Prabhat

(c) Mukta (d) Sharda

40. Which pulse has lowest harvesting index?

(a) Bengal gram (b) Green gram

(c) Red gram (d) Kabuli

41. 1st interspecific hybrid of cotton is

(a) Savita (b) Surya

(c) H4 (d) Varalaxmi

42. Fibre maturity of cotton is measured by

(a) Tendrometer (b) Arealometer

(c) Nepiness (d) Hardiness

43. Optimum temperature for retting of jute is around (in degree Celsius)

(a) 28 (b) 34

(c) 40 (d) 25

44. Hollow heart in potato cause due to

(a) Lack of oxygen (b) Excess of oxygen

(c) Lack of Nitrogen (d) Excess of Nitrogen

45. Hulling percentage of rice and shelling percentage of groundnut are respectively (in percentage)

(a) 66, 70 (b) 70, 66

(c) 88, 92 (d) None of the above

46. What is the existing annual income limit for identification of economically weaker section (EWS)?

(a) 8 lakhs (b) 10 lakhs

(c) 6 lakhs (d) 4 lakhs

47. Which is India's first certified organic fruit?

(a) Apple (b) Strawberry

(c) Kiwi (d) Banana

48 which Indian has been named as CEO of twitter recently

(a) Arvind Krishna (b) Shantanu Narayana

(c) Parag Agrawal (d) Shahrukh khan

49. head quarter of international solar alliance is

(a) Gurugram (b) Delhi

(c) Colombo (d) Moscow

50. Best actor at BRICS film festival recently

(a) Dhanush (b) Amitabh Bachhan

(c) Ranbir Kapoor (d) Ranveer Singh

ANSWER KEY

1	2	3	4	5	6	7	8	9	10	11	12	13	14	15
(b)	(b)	(b)	(a)	(a)	(c)	(a)	(c)	(b)	(b)	(b)	(b)	(c)	(c)	(c)
16	17	18	19	20	21	22	23	24	25	26	27	28	29	30
(d)	(a)	(a)	(c)	(c)	(a)	(c)	(b)	(d)	(d)	(b)	(c)	(c)	(b)	(b)
31	32	33	34	35	36	37	38	39	40	41	42	43	44	45
(b)	(c)	(c)	(b)	(a)	(d)	(b)	(a)	(b)	(c)	(d)	(b)	(b)	(d)	(a)
46	47	48	49	50										
(a)	(c)	(c)	(a)	(a)										

Chapter 4

Crop Production of Cotton, Jute, Potato, Groundnut, Soyabean, Mustard, Tea, Tobacco, Buck Wheat

1. **World's first Nano urea liquid developed by-**
   (a) NFL (b) IFFCO
   (c) IARI (d) CFQCTI

2. **The country with highest number of organic producers in the world-**
   (a) India (b) Australia
   (c) Kenya (d) Bhutan

3. **World earliest maturity variety of cotton genotype was identified by CICR scientist (2017) -Variety name?**
   (a) Maru Vikas (b) Varalaxmi
   (c) Surya (d) Yugank

4. **First country to approve GM Wheat crop (variety HB4) tolerant to drought and ammonium glufosinate herbicide is?**
   (a) Argentina (b) India
   (c) Switzerland (d) Germany

5. **GM Corn variety developed by China to fight against invasive Fall Army Worm is-**
   (a) 005 (b) 008
   (c) 007 (d) 001

6. **Which of the following is known as pseudo cereal?**
   (a) Wheat (b) Sorghum
   (c) Napier grass (d) Buck wheat

7. **Tokyo is a variety of**
   (a) Pigeon pea (b) Tobacco
   (c) Potato (d) Buck wheat

8. **Fruit of buck wheat is known as**
   (a) Drupe (b) Berry
   (c) Achene (d) Pome

9. **The alkaloid named rutin present in buckwheat is present in**
   (a) Leaves (b) Flower
   (c) Root (d) Both (a) and (b)

10. **Family of buck wheat is**
   (a) Graminae (b) Malvaceae
   (c) Polygonaceae (d) Solanaceae

11. **Removal of matured leaves in Tobacco is known as?**
   (a) Topping (b) Desuckering
   (c) Priming (d) Curing

12. **Removal of flower heads with few upper or top leaves to improve the plant size and quality is known as?**
   (a) Topping (b) Desuckering
   (c) Priming (d) Curing

13. **Removal of axillary buds to divert energy and nutrients from flower head to leaves is known as?**
   (a) Topping (b) Desuckering
   (c) Priming (d) Curing

14. **The moisture removing drying process to impart required colour, texture and aroma to the final product is known as?**
   (a) Topping (b) Desuckering
   (c) Priming (d) Curing

15. **Which of the following belongs to FCV type of tobacco?**
   (a) Bidi tobacco (b) Hookah tobacco
   (c) Chewing tobacco (d) Cigarette tobacco

16. First hybrid variety of tobacco in India is

(a) GTV-1 (b) GTH-1
(c) HB-1 (d) Pusa tobacco hybrid-1

17. Which of following type of tobacco belongs to Non Virginia type tobacco?

(a) Bidi tobacco (b) Hookah tobacco
(c) Chewing tobacco (d) All of the above

18. Nicotine content is more in

(a) *N. tabacum* (b) *N. rustica*
(c) Both contain same amount (d) None of the above

19. Queen of beverage is?

(a) Coffee (b) Tea
(c) Sarbat (d) Dalgona coffee

20. Which country is the largest producer, consumer and exporter of tea in the world?

(a) India (b) Sri Lanka
(c) Denmark (d) Japan

21. The best method of plucking of tea

(a) 1 leaf 1 bud (b) 2 leaf 1 bud
(c) Only 2 leaves (d) 3 leaves 1 bud

22. Which state stands 1st position in tea production in India?

(a) Karnataka (b) West Bengal
(c) Kerala (d) Assam

23. Pungency of mustard is due to?

(a) Isothiocyanate (b) 2 methyl cyanate
(c) Allyl iso fluvalinate (d) Isobutane

24. Pusa Jaikishan is developed by the technique?

(a) Protoplasmic fusion (b) Somaclonal variation
(c) Somatic hybridization (d) None of the above

25. First mustard variety in the world is?

(a) NRCHB-506 (b) NRCHB-101
(c) Pusa bold (d) Pusa jai kisan

26. First hybrid of mustard in India is

(a) Pusa jai kisan (b) NRCHB-506

(c) NRCHB-101 (d) Kranti

27. Which crop is known as wonder crop or yellow jewel?

(a) Cotton (b) Buckwheat

(c) Soyabean (d) Tobacco

28. Which is known as poor man's crop and rich man's food

(a) Lentil (b) Soyabean

(c) Drumstick (d) Red gram

29. Bragg, Lee, Kent, Clark are the varieties of which crop?

(a) Tea (b) Cotton

(c) Soyabean (d) Tobacco

30. Which is known as poor man's meat

(a) Soyabean (b) Lentil

(c) Both of the above (d) None of the above

31. Largest groundnut producing state is?

(a) Gujarat (b) Tamil Nadu

(c) Odisha (d) Maharashtra

32. The groundnut rosette virus is transmitted by which insect

(a) Thrips (b) Aphids

(c) Leaf hopper (d) Tree hopper

33. Nitrogen percentage in groundnut cake is

(a) 3.9 per cent (b) 3.7 per cent

(c) 7.3 per cent (d) 9.3 per cent

34. Which of the following is the example of non-edible oil cake?

(a) Groundnut cake (b) Linseed cake

(c) Castor cake (d) Rapeseed cake

35. Pegging stage of groundnut comes after

(a) 30 DAS (b) 45 DAS

(c) 55 DAS (d) 65 DAS

36. Aflatoxin is produced by

(a) *Aspergillus niger*
(b) *Aspergillus flavus*
(c) *Aspergillus fumigatus*
(d) *Aspergillus terraeus*

37. Popping in groundnut is caused due to

(a) Calcium deficiency
(b) Calcium excesses
(c) Fe deficiency
(d) Fe excesses

38. Peg is modification of

(a) Gynophores
(b) Androphores
(c) Androgynophores
(d) None of the above

39. Oil and protein contents in groundnut (in percentage)

(a) 26,45
(b) 45,26
(c) 40,20
(d) 20,40

40. High yielding type of groundnut is

(a) Spreading type
(b) Bunch type
(c) Erect type
(d) None of the above

41. Black heart disease of potato is caused by?

(a) Lack of oxygen
(b) Excess of oxygen
(c) Excess of nitrogen
(d) Lack of nitrogen

42. Ribboning is a process done in?

(a) Tea
(b) Jute
(c) Cotton
(d) Potato

43. For obtaining quality fibre and good yield jute should be harvested at?

(a) Small pod stage
(b) Large pod stage
(c) At any stage
(d) 2 leaf stage

44. The glycoside present in white/tita jute is known as?

(a) Curcumin
(b) Corcharine
(c) Both
(d) None of the above

45. 1 bale of cotton, jute, mesta and silk (national) is respectively? (In kg)

(a) 180, 181, 170 50
(b) 181, 180, 50, 170
(c) 170, 180, 181, 50
(d) 50, 170, 180, 181

46. Lint percentage in cotton is? (In per cent)

(a) 50
(b) 66
(c) 33
(d) 24

47. Fibre thickness of cotton is measured by?

(a) Number of counts
(b) Arealometer
(c) Nepiness
(d) In bale

48. Seed rate (kg/ha) and average plant population per ha of Bt cotton is?

(a) 15-20, 100000 plants
(b) 2-3, 66000
(c) 1-1.5, 10000
(d) 10-15, 80000

49. Which country recently amends national anthem to honour indigenous people

(a) India
(b) Australia
(c) Germany
(d) Japan

50. India's highest meteorological centre has been inaugurated in which state/UT?

(a) Sikkim
(b) Ladakh
(c) Madhya Pradesh
(d) Uttar Pradesh

ANSWER KEY

1	2	3	4	5	6	7	8	9	10	11	12	13	14	15
(b)	(a)	(d)	(a)	(c)	(d)	(d)	(c)	(d)	(c)	(c)	(a)	(b)	(d)	(d)
16	17	18	19	20	21	22	23	24	25	26	27	28	29	30
(b)	(d)	(b)	(b)	(a)	(b)	(d)	(a)	(c)	(d)	(b)	(c)	(c)	(c)	(c)
31	32	33	34	35	36	37	38	39	40	41	42	43	44	45
(a)	(b)	(c)	(c)	(c)	(b)	(a)	(a)	(b)	(a)	(a)	(b)	(a)	(b)	(c)
46	47	48	49	50										
(c)	(c)	(c)	(b)	(b)										

Chapter 5

Weed Science, Weed Dormancy, Bioherbicides, Biological Control of Weeds, Herbicides, Fertilizers, other Important Terms

1. **Which country was the host country for the 2021 world environment day?**
   (a) Qatar (b) India
   (c) Afghanistan (d) Pakistan
2. **Which state of India has been invited to participate in international seed conference conducted by FAO?**
   (a) Tamil Nadu (b) Telangana
   (c) Kerala (d) Andhra Pradesh
3. **India's 1st food museum has been launched in which state?**
   (a) Tamil Nadu (b) Kerla
   (c) Odisha (d) Uttarakhand
4. **India's first grass conservatory "Germplasm conservation centre" has been inaugurated in which state?**
   (a) Himachal Pradesh (b) Uttar Pradesh
   (c) Uttarakhand (d) Haryana
5. **Which state has become the first one to impose stock limits on edible oil?**
   (a) Karnataka (b) Madhya Pradesh
   (c) Uttar Pradesh (d) Andhra Pradesh

6. All Three types (Enforced, innate, induced) of weed dormancy is seen in

(a) Wild oat (b) Cynodon
(c) Cyprus (d) Canary grass

7. A weed with a balloon structure for effective dissemination

(a) *Physalis minima* (b) *Avena fatua*
(c) *Phalaris minor* (d) *Amaranthus viridis*

8. A sedge with rhizomes

(a) *Commelina obliqua* (b) *Cyperus rotundus*
(c) *Scirpus* sp. (d) *Eleocharis*

9. Hans Molish is associated with

(a) Competition (b) Eutrophication
(c) Allelomediation (d) Allelopathy

10. The partial root parasite

(a) Cuscuta (b) Loranthus
(c) Striga (d) Orobanche

11. The most systematic method for classifying weeds is based on

(a) Morphology (b) Life history
(c) Habitat (d) Phylogenetic

12. Bioagent for controlling Parthenium hysterophorus

(a) *Bactra verutana* (b) *Zygogramma bicolorata*
(c) Chrysolira (d) Flea beetle

13. Dredging is most effective treatment for controlling

(a) Parasitic weeds (b) Lantana
(c) Agrestals (d) Aquatic weeds

14. Flooding – a worldwide crop husbandry method of controlling weeds in

(a) Maize (b) Sugarcane
(c) Buckwheat (d) Rice

15. A Natural amongst the following

(a) Devine (b) Glyphosate
(c) Potassium chloride (d) 2,4-D

16. The first chemical shown to have herbicidal activity

(a) 2,4-D (b) Bordeaux mixture

(c) Paraquat (d) 2,4- DB

17. The herbicidal activity of 2,4-D at the first time was reported by

(a) Hammer and Tukey (1944)

(b) Zimmerman and Hitchcock (1942)

(c) Nutman and Blackman (1945)

(d) None of the above

18. A practice where one or two flushes of weeds are destroyed before seeding a crop

(a) Cheeling (b) Stale seed bed

(c) Searing (d) None of the above

19. In case of combined application of herbicides, application rate can be reduced in which of the following interaction

(a) Antagonistic (b) Synergistic

(c) Additive (d) None of the above

20. Test weight of Phalaris minor is

(a) 10 g (b) 15 g

(c) 7g (d) 2 g

21. Weed responsible for forest fire is

(a) Aligator weed (b) Sorghum halepense

(c) Lantana (d) Amaranthus

22. Which herbicide is systemic and translocated type?

(a) Atrazine (b) Diuron

(c) Paraquat (d) Glyphosate

23. A pre plant incorporated herbicide

(a) Paraquat (b) Delapon

(c) Uracils (d) Fluchloralin

24. Trade name of Nitrofen?

(a) Lasso (b) Toke - E - 25

(c) Round up (d) Plant guard

25. Paraquat herbicide belong to which group

(a) Bipyridylium
(b) Benzoic
(c) Carboxalic
(d) Uraciles

26. Which type of clay will adsorb more herbicide?

(a) Montmorillonite
(b) Illite
(c) Kaolinite
(d) Both (b) and (c)

27. The type of soils which require higher dose of herbicides to be effective is

(a) Inceptisols
(b) Histosols
(c) Oxisols
(d) Ultisols

28. The potential of a pesticide for causing damage to plants is its:

(a) Phytotoxicity
(b) Lethal dose
(c) Defoliation ability
(d) Chronicity

29. The authority associated with the first genetically modified plants the soybean resistant to glyphosate from Monsanto which achieved widespread success was

(a) Rachel Carson (27 May, 1907–14 April, 1964)
(b) Fritz Haber (9 December 1868–29 January 1934)
(c) Robert Fraley (birth date: 1953)
(d) Norman Borlaug (25 March, 1914–12 September, 2009)

30. The first genetically modified plants, the soybean resistant to glyphosate was developed during

(a) 1983
(b) 2003
(c) 1993
(d) 1973

31. Herbicide Antidotes

(a) Protect crop plants
(b) Harm crop plants
(c) Protect weeds
(d) Harm weeds

32. What herbicide classification does 2-4-D fall into?

(a) Soil, Short residual, non-selective
(b) Foliage, contact, non-selective
(c) Foliage, translocative, selective
(d) Soil, Long residual, selective

33. A herbicide applied to the soil before planting a crop is called a

(a) Pre-emergence
(b) Post-emergence
(c) Pre-plant
(d) None of the above

34. Herbicide are registered under which insecticide registration act

(a) 1968 (b) 1969

(c) 1964 (d) 1962

35. Glufosinate (Bar) is resistant trait of which crop

(a) Soybean (b) Sugar beet

(c) Maize (d) Cotton

36. First mycoherbicide developed by Abbott Laboratories, USA was

(a) Biosedge (b) Devine

(c) Collego (d) Biomal

37. Which crop is an indicator plant for the bio assay of Triazines and Aliphatics?

(a) Sugar beet (b) Sorghum

(c) Cotton (d) Mustard

38. ______ is a selective post-emergence herbicide used for weed control in rice

(a) Bispyribac-sodium (b) Butachlor

(c) Pretilachlor (d) Atrazine

39. Weed intensity if number of weeds is110/m2 and wheat plants are 90/ m[57].

(a) 55 per cent (b) 110 per cent

(c) 22.5 per cent (d) 77.5 per cent

40. First introduced organic herbicide in the world is

(a) Dinoseb (b) 2,4-D

(c) MCPA (d) All of the above

41. An available soil nutrient determined in terms of a standard fertilizer used is known as

(a) A-value (b) Z- value

(c) F- value (d) D-value

42. Which is the contact, non-selective and zero persistent herbicide in soils?

(a) Paraquat (b) Atrazine

(c) 2,4-D (d) Bromacil

43. If the yield from weed free is 3500 kg and that from weedy 1500 kg/ha, weed index would be

(a) 57.1 per cent
(b) 133.3 per cent
(c) 45.0 per cent
(d) 25.6 per cent

44. A ceremonial weed

(a) *Asparagus racemosus* (Shatawari)
(b) *Cynodon dactylon* (Dhoob ghaas)
(c) *Nelumbo nucifera* (Lotus flower)
(d) All of the above

45. Out of 50 quadrats studied, if Echinochloa is found in 26 quadrats, its frequency of occurrence would be

(a) 13 per cent
(b) 26 per cent
(c) 52 per cent
(d) 68 per cent

46. Recently 2022, At India gate which freedom fighter's grand granite statue will be installed

(a) Bhagat Singh
(b) Uddham Singh
(c) Sardar Vallabhbhai Patel
(d) Netaji Subash Chandra Bose

47. Which Asian Country recently unveiled its National Security Policy (NSP)?

(a) Sri Lanka
(b) Pakistan
(c) Afghanistan
(d) Bangladesh

48. What is the limit of offline digital payment per transaction, as per RBI's recent notification?

(a) Rs 100
(b) Rs 200
(c) Rs 500
(d) Rs 1000

49. India's first multi-dimensional 'adventure sports expedition' was carried out in which country?

(a) USA
(b) France
(c) Switzerland
(d) Australia

50. Which film won the 'Best picture (drama)' award at the Golden Globes 2022?

(a) Pushpa
(b) Atrangi re
(c) The Power of a Dog
(d) King Richard

ANSWER KEY

1	2	3	4	5	6	7	8	9	10	11	12	13	14	15
(d)	(b)	(a)	(c)	(c)	(a)	(a)	(b)	(d)	(c)	(d)	(b)	(d)	(d)	(a)
16	17	18	19	20	21	22	23	24	25	26	27	28	29	30
(b)	(b)	(b)	(b)	(d)	(c)	(d)	(d)	(b)	(a)	(a)	(b)	(a)	(c)	(a)
31	32	33	34	35	36	37	38	39	40	41	42	43	44	45
(a)	(c)	(c)	(a)	(d)	(b)	(a)	(a)	(a)	(a)	(a)	(a)	(a)	(d)	(c)
46	47	48	49	50										
(d)	(b)	(b)	(b)	(c)										

SOIL SCIENCE

Chapter 6

Introduction of Soil Science, Types of Rocks, Soil Formation, Components of Soil, Soil Profile, Properties of Soil, Soil Structure and Texture, Types of Soil Structure, Factors Affecting Soil Structure

1. **Which is the only city of India to be recognised as "Tree city of the world" by FAO recently**
   (a) Bhubaneshwar (b) Hyderabad
   (c) Chennai (d) Bangalore
2. **World's largest solar tree is installed at**
   (a) Odisha (b) Haryana
   (c) Bihar (d) West Bengal
3. **World's first conservation area for sea cucumber created at?**
   (a) Lakshadweep (b) Chandigarh
   (c) Switzerland (d) London
4. **International rice research centre (IRRI) South Asia regional centre established in 2018 at**
   (a) Kathmandu (b) Varanasi
   (c) Chandigarh (d) Bhubaneshwar
5. **First state in India to start Automated Weather station?**
   (a) Maharashtra (b) Goa
   (c) Punjab (d) New Delhi

6. **______ is the category of igneous rock?**
   (a) Granite rocks (b) Basalt
   (c) Both (a) and (b) (d) Quartz

7. **______ is the category of sedimentary rock?**
   (a) Lime stone (b) Sandstone
   (c) Dolomite (d) All of the above

8. **______ is the category of metamorphic rock?**
   (a) Gneiss (b) Quartz
   (c) Marble (d) All of the above

9. **Percentage of feldspars in earth crust is.?**
   (a) 48 per cent (b) 38 per cent
   (c) 58 per cent (d) 28 per cent

10. **Father of soil science and pedology?**
   (a) Dokuchev (b) Traug
   (c) Winogradski (d) None of the above

11. **Quartz constitutes how much percent of earth crust?**
   (a) 36 (b) 48
   (c) 56 (d) 60

12. **Study of various property of soil in relation to growth, nutrition and yield of crops?**
   (a) Pedology (b) Edaphology
   (c) Soil Chemistry (d) Soil Physics

13. **Rate according to the decreasing order of resistant to weathering**
   (a) Quartz>Fledspar>black mica>white mica
   (b) Quartz>Black mica>White mica>Fledspar
   (c) Quartz>White mica>Black mica>Fledspar
   (d) Fledspar>Black mica>White mica>Quartz

14. **Quartz is present in a pure form as a prominent constituent in?**
   (a) Acid rock (b) Basic rock
   (c) Alkali rock (d) Same in All of the above

15. **The hardest rock is**
   (a) Diamond (b) Talc
   (c) Quartz (d) Marble

16. **Potash mica is**
   (a) Black mica (b) Biotite
   (c) Muscovite (d) Both (a) and (b)

17. **Partially decomposed organic material is known as?**
   (a) Peat soil (b) Muck soil
   (c) Sandy soil (d) Light soil

18. **Magnesium mica is**
   (a) Muscovite (b) Biotite
   (c) Black mica (d) Both (b) and (c)

19. **Which of the following is the active factor for soil formation?**
   (a) Parent material (b) Topography
   (c) Climate (d) Age of land

20. **The vertical section of the soil through all the horizon is known as?**
   (a) Soil horizon (b) Soil profile
   (c) Soil divider (d) Soil formation

21. **Top most mineral horizon is**
   (a) O (b) A
   (c) B (d) C

22. **Zone of washing out or zone of leaching out**
   (a) O (b) A
   (c) B (d) C

23. **Zone of washing in is known as?**
   (a) O (b) A
   (c) B (d) C

24. **The horizon referred as subsoil?**
   (a) O (b) A
   (c) B (d) C

25. **Zone of least weathering is known as?**
   (a) O (b) A
   (c) B (d) C

26. **Organic layer is**
   (a) O (b) A
   (c) B (d) C

27. Solum is formed by which soil horizons

(a) O+A
(b) A+B
(c) B+C
(d) C+R

28. The highest proportion (Component) of soil

(a) Mineral
(b) Organic matter
(c) Air space
(d) Water

29. Relative proportions of various types of soil particles is known as?

(a) Soil texture
(b) Soil structure
(c) Bulk density
(d) Particle density

30. The soil texture is measured by

(a) Feel method
(b) International pipette method
(c) Bouyoucos hydrometer method
(d) All of the above

31. Highest pore space present in

(a) Sandy soil
(b) Silt soil
(c) Clay soil
(d) None of the above

32. Sandy soil contain more than how much percent of sand

(a) 85
(b) 80
(c) 40
(d) None of the above

33. Clay soil contain more than how much percent of clay?

(a) 85
(b) 80
(c) 40
(d) None of the above

34. Which soil is best for agricultural production?

(a) Clay
(b) Sandy
(c) Sandy loam
(d) loam

35. The arrangement of the soil particle is known as

(a) Soil structure
(b) Soil texture
(c) particle density
(d) Bulk density

36. Which ion destroy the soil structure due to deflocculation?

(a) Sodium
(b) Calcium
(c) Magnesium
(d) Silicon

37. Which ion destroy the soil due to flocculation?

(a) Ca (b) Sodium

(c) Chloride (d) Potassium

38. Soil structure is divided into different types based on?

(a) Shape or form and arrangement pattern of peds

(b) Size of peds

(c) dDegree of distinctness

(d) Coherent force

39. Soil structure is divided into different class based on?

(a) Shape or form and arrangement patterns of peds

(b) Size of peds

(c) Degree of distinctness

(d) Coherent force

40. Soil Structure is divided into different grades based on?

(a) Shape or form and arrangement patterns of peds

(b) Size of peds

(c) Degree of distinctness

(d) Coherent force

41. Stability of the aggregates is determined by

(a) Types of soil structure (b) Class of soil structure

(c) Grade of soil structure (d) None of the above

42. Coherent mass of soil broken into any shape by artificial means such as tillage?

(a) Clod (b) Peds

(c) Texture (d) Structure

43. Natural aggregates of soil is known as

(a) Structure (b) Texture

(c) Peds (d) Clod

44. Which property of the soil can't be changed?

(a) Soil texture (b) Soil structure

(c) Soil density (d) Thermal conductivity

45. According to ISSS, Diameter <0.002 is classified under

(a) Gravel (b) Sand

(c) Silt (d) Clay

46. 'The 2021 Yidan Prize' which was awarded recently, is associated with which field?

(a) Business (b) Sports

(c) Science and Technology (d) Education

47. Who was Google's most searched-for personality in India in 2021?

(a) Neeraj Chopra (b) Elon Musk

(c) P V Sindhu (d) Mirabai Chan

48. As per recent research from IISc, which chemical compound can suppress HIV?

(a) Hydrogen sulphide (b) Nitrous Oxide

(c) Sulphur Di oxide (d) Carbon Di sulphide

49. Which state topped the "Good Governance Index 2021"?

(a) Tamil Nadu (b) Gujarat

(c) Kerala (d) Andhra Pradesh

50. What is the GST to be collected by Food aggregators like Swiggy and Zomato from January 1, 2022?

(a) 5 per cent (b) 8 per cent

(c) 12 per cent (d) 18 per cent

ANSWER KEY

1	2	3	4	5	6	7	8	9	10	11	12	13	14	15
(b)	(d)	(a)	(b)	(a)	(c)	(d)	(d)	(a)	(a)	(a)	(b)	(c)	(a)	(a)
16	17	18	19	20	21	22	23	24	25	26	27	28	29	30
(c)	(a)	(d)	(c)	(b)	(b)	(b)	(c)	(c)	(d)	(a)	(b)	(a)	(a)	(d)
31	32	33	34	35	36	37	38	39	40	41	42	43	44	45
(c)	(a)	(c)	(c)	(a)	(a)	(a)	(a)	(b)	(c)	(c)	(a)	(c)	(a)	(d)
46	47	48	49	50										
(d)	(a)	(a)	(b)	(a)										

Chapter 7

Soil Density, Thermal Conductivity, Soil Plasticity, Soil Consistency, Soil Colour, Soil Porosity, Silicate Minerals, Classification of Silicate Clays

1. Per ha consumption of fertilizers Lowest in

(a) Tamil Nadu
(b) Kerala
(c) Punjab
(d) Odisha

2. Krishi Karman Award for the highest production of food grain (>10 mt) awarded in 2021 to?

(a) Chhattisgarh
(b) Bihar
(c) Uttar Pradesh
(d) Punjab

3. First institute in the world that is exclusively devoted to gender related research in agriculture?

(a) CIWA-Bhubaneshwar
(b) CICR-Nagpur
(c) NIBSM
(d) NIASM

4. "Virtual Classroom" and Web Education Channel" of ICAR-NAHEP is?

(a) Agri siksha
(b) Agri diksha
(c) Agri vidya
(d) None of the above

5. No of pesticides registered by 2022 is

(a) 275
(b) 292
(c) 315
(d) 410

6. **Soil Density is measured by**
   (a) Hygrometer (b) Pycnometer
   (c) piezometer (d) Densometer

7. **Weight of individual soil particle per unit volume is called**
   (a) Bulk density (b) Particle density
   (c) Soil consistency (d) ODR

8. **Particle density of mineral soil? (g/cc)**
   (a) 2.66 (b) 2.3
   (c) 1.5 (d) 1

9. **Particle density of Organic soil? (g/cc)**
   (a) 2.1-2.4 (b) 1.5
   (c) 1 (d) 2.66

10. **Particle density is also known as?**
   (a) Real density (b) True density
   (c) Both (a) and (b) (d) Pseudo density

11. **Particle density of organic matter?**
   (a) 2.66 (b) 1
   (c) 1.1-1.4 (d) 2.1-2.4

12. **By addition of Organic matter and mineral what changes in particle density respectively?**
   (a) Decrease, Increase (b) Increase, Decrease
   (c) Decrease, Decrease (d) Increase, Increase

13. **Weight of oven dried soil to the bulk volume including pore space is known as**
   (a) Particle density (b) Bulk density
   (c) Apparent density (d) Both (b) and (c)

14. **Which statement is correct**
   (a) Bulk density is less than particle density
   (b) Particle density excludes the porespace
   (c) Bulk density is half of particle density
   (d) All of the above

15. Bulk density of mineral soil is?

(a) 1.33
(b) 2.66
(c) 1.8-1.9
(d) 3.2

16. Bulk density of organic matter?

(a) 0.5
(b) 1.5
(c) 2
(d) 2.5

17. Bulk density of compact subsoil?

(a) 1
(b) 2
(c) 1.5
(d) 1.2

18. Effect of compactness on soil bulk density?

(a) Increase
(b) Decrease
(c) Same
(d) No effect

19. Soil with large pores effects on bulk density?

(a) Increase
(b) Lower
(c) Constant
(d) None of the above

20. Which statements are true

(a) Sandy soil have higher bulk density than clay
(b) Sandy soil have higher bulk density than loam
(c) Organic matter addition decreases bulk density
(d) All of the above

21. Amount of heat flowing in unit time through a cross section of soil is known as?

(a) Thermal conductivity
(b) ODR
(c) Consistency
(d) Plasticity

22. High thermal conductivity in?

(a) Sand
(b) Loam
(c) Clay
(d) All same

23. Effect of Soil moisture and organic matter on soil thermal conductivity?

(a) Increase, Decrease
(b) Decrease, Decrease
(c) Increase, Increase
(d) Decrease, Increase

24. Effect of Porosity and tillage on thermal conductivity?

(a) Increase, Decrease
(b) Decrease, Decrease
(c) Increase, Increase
(d) Decrease, Increase

25. Redox potential for well-drained soil?

(a) 400-700mv (b) -250 to -350 mv

(c) 200-400 mv (d) -150 to -200 mv

26. Redox potential of water-logged soil is?

(a) 400-700mv (b) -250 to -350 mv

(c) 200-400 mv (d) -150 to -200 mv

27. Plasticity of soil is observed when the percentage of clay is more than

(a) 5 per cent (b) 10 per cent

(c) 15 per cent (d) None of the above

28. Soil plasticity is determined by

(a) Pycnometer (b) Liquid limit device

(c) Piezometer (d) Boyoucous hydrometer

29. Plasticity index is given by

(a) Rouxberg (b) Alterberg

(c) Dokuchev (d) Winogradsky

30. Lowest moisture content at which soil can be deformed without cracking?

(a) Lowest plastic limit (b) Upper limit

(c) Liquid limit (d) Both (b) and (c)

31. Higher moisture content at which destroys soil aggregate

(a) Lowest plastic limit (b) Upper limit

(c) Liquid limit (d) Both (b) and (c)

32. The soil consistency is correct in which option

(a) Dry>moist>wet (b) Moist>dry>wet

(c) Wet>dry>moist (d) Wet=dry=moist

33. ODR is measured by

(a) Platinum weir (b) Boucous hydrometer

(c) International pipete method (d) Pycnometer

34. Which of the following represents dominant spectral colour?

(a) Hue (b) Value

(c) Chroma (d) None of the above

35. Lightness or darkness of colour is represented by?

(a) Hue (b) Value

(c) Chroma (d) All of the above

36. Purity of colour is measured by

(a) Hue (b) Value

(c) Chroma (d) All of the above

37. Specific heat of soil?

(a) 1 (b) 0.2

(c) 0.5 (d) None of the above

38. Macropores are found more in?

(a) Sandy soil (b) Loamy soil

(c) Clay soil (d) None of the above

39. Micropores are found more in

(a) Sandy soil (b) Loamy soil

(c) Clay soil (d) None of the above

40. Which of the following is boron alumina silicate

(a) Tourmaline (b) Apatite

(c) Biotite (d) Dolomite

41. Which of the silicate contain Potassium

(a) Biotite (b) Olivine

(c) Orthoclase (d) Both (a) and (c)

42. Olivine contains

(a) Mg (b) Mo

(c) K (d) Both (a) and (b)

43. Which element composition is highest in earth crust?

(a) Oxygen (b) Silicon

(c) Aluminium (d) Ferrous

44. 1:1 type mineral is

(a) Kaolinite (b) Brucite

(c) Montmorillonite (d) None of the above

45. Highest CEC is observed in

(a) Kaolinite (b) Illite

(c) Vermiculite (d) Humus

46. Which is the most innovative educational institute of India in the technical category, as per the ARIIA 2021 ranking?

(a) IIT Kanpur
(b) IIT Madras
(c) IIT, DELHI
(d) AIIMS, DELHI

47. Which Indian state recorded the highest Tiger deaths in 2021?

(a) Madhya Pradesh
(b) Uttar Pradesh
(c) Andhra Pradesh
(d) West Bengal

48. As on January 1, 2022, what is the estimated population of the world, as per the US Census Bureau?

(a) 3.8 billion
(b) 5.8 billion
(c) 7.8 billion
(d) 9.8 billion

49. Which Indian state is set to host the next edition of Khelo India Games in 2023?

(a) Odisha
(b) Karnataka
(c) Andhra Pradesh
(d) Uttar Pradesh

50. June 14, 2020, famous Bollywood actor Sushant Singh Rajput passed away. Which was his first film in a leading role?

(a) Raabta
(b) Kedarnath
(c) Kai Po che
(d) MS Dhoni-the untold story

ANSWER KEY

1	2	3	4	5	6	7	8	9	10	11	12	13	14	15
(b)	(b)	(a)	(b)	(b)	(b)	(b)	(a)	(a)	(c)	(c)	(a)	(d)	(d)	(a)
16	17	18	19	20	21	22	23	24	25	26	27	28	29	30
(a)	(b)	(a)	(b)	(d)	(a)	(a)	(a)	(b)	(a)	(b)	(c)	(b)	(b)	(a)
31	32	33	34	35	36	37	38	39	40	41	42	43	44	45
(d)	(a)	(a)	(a)	(b)	(c)	(b)	(a)	(c)	(a)	(d)	(d)	(a)	(a)	(d)
46	47	48	49	50										
(b)	(a)	(c)	(b)	(c)										

Chapter 8

Modern Soil Classification, Land Capability Classification, Soils of India, Soil Fertility and Soil Productivity, Humus, Organic Matter, Flocculation, Deflocculation, Soil Buffering, Saline Soil, Alkaline Soil, Problematic Soil and Reclamation of Soil

1. **Which state has set to become *Har Ghar Jal* state by 2022?**
   (a) Manipur (b) Assam
   (c) Punjab (d) Telangana
2. **Central govt has removed the ban on which product exports from 2021?**
   (a) Rose (b) Onion
   (c) Garlic (d) Oil
3. **Rice ATM to be set up in state**
   (a) Karnataka (b) Gujarat
   (c) West Bengal (d) Odisha
4. **Which state launched Kishan Kalyan Mission?**
   (a) Madhya Pradesh (b) Uttar Pradesh
   (c) Gujarat (d) Maharashtra
5. **Krishi sanjeevani van started in**
   (a) Karnataka (b) Uttar Pradesh
   (c) Madhya Pradesh (d) Andhra Pradesh

6. **C:N ratio of humus is**
   (a) 10:1 (b) 12:1
   (c) 5:1 (d) 20:1

7. **The ability of the soil to provide all the essential nutrients in available form is known as?**
   (a) Soil Fertility (b) Soil productivity
   (c) Soil buffering (d) None of the above

8. **Capability of soil to produce specified crop yield under well-defined and specified system of management of inputs and environment condition is known as**
   (a) Soil fertility (b) Soil productivity
   (c) Soil buffering (d) None of the above

9. **Which of the following is the inherent property of soil**
   (a) Soil fertility (b) Soil productivity
   (c) Soil buffering (d) None of the above

10. **Organic matter=Organic carbon x?**
    (a) 1.456 (b) 1.732
    (c) 1.896 (d) 1.372

11. **Management of Acidic soil is done by application of?**
    (a) Rock phosphate (b) Press mud
    (c) Gypsum (d) Both (a) and (b)

12. **Management of saline soil is done by?**
    (a) Application of rock phosphate (b) Flooding
    (c) FYM (d) Both (b) and (c)

13. **Reclamation of alkali soil is done by application of?**
    (a) Gypsum (b) Iron pyrite
    (c) Green Manuring (d) All of the above

14. **Among fruits crops which of the following is highest salt tolerant crop?**
    (a) Date palm (b) Phalsa
    (c) Mango (d) Pomegranate

15. **Among field crops highest salt tolerant crop is?**
    (a) Barley (b) Tobacco
    (c) Cotton (d) Mustard

16. Availability of P is maximum at pH?

(a) <6.0
(b) 6.0-6.5
(c) >7
(d) All of the above

17. Ca and Mo availability is maximum at pH?

(a) <6.0
(b) 6.0-6.5
(c) >7
(d) All of the above

18. Al and Fe availability is maximum at pH?

(a) <6.0
(b) 6.0-6.5
(c) >7.0
(d) All of the above

19. Which of the following ion has highest flocculation capacity?

(a) Al
(b) Ca
(c) Mg
(d) H

20. Saline soil is also known as?

(a) White alkali
(b) Solon chalk
(c) Sodic
(d) Both (a) and (b)

21. which is the largest soil order in the world?

(a) Inseptisol
(b) Entisol
(c) Vertisol
(d) Oxisol

22. Self-mulching is occurred in the soil order?

(a) Entisol
(b) Vertisol
(c) Oxisol
(d) alfisol

23. Peat and muck soil is observed in which order?

(a) Oxisols
(b) Alfisols
(c) Histosols
(d) Aridisols

24. Cryoturbation is occurred in which soil order?

(a) Andisols
(b) Gellisols
(c) Mollisols
(d) Ultisols

25. Volcanic ash is formed by which soil order?

(a) Ultisol
(b) Andisol
(c) Gellisol
(d) Mollisols

26. Which soil order is known as "under ash?"

(a) Spodosols (b) Ultisols
(c) Aridisols (d) Andisols

27. US soil conservation service divided land capability classes into how many groups?

(a) 8 (b) 12
(c) 16 (d) 5

28. Alluvial soil and Indo-Gangetic plains comes under which land capability class?

(a) I (b) II
(c) III (d) IV

29. Land used for Pasture and hay comes under which land capability class?

(a) I (b) II
(c) III (d) IV

30. Major group of soil present in India?

(a) Alluvial (b) Black
(c) Red (d) Laterite

31. Khadar, Bhangar, Tarai and bhagar belongs to which soil group?

(a) Alluvial soil (b) Black soil
(c) Red soil (d) Laterite soil

32. Regur belongs to which group of soil?

(a) Alluvial (b) Black cotton
(c) Red (d) Desert

33. Red soil is present majorly in which state

(a) Tamil Nadu (b) Andhra Pradesh
(c) Uttar Pradesh (d) Madhya Pradesh

34. Maximum leaching occurs in which soil group?

(a) Red (b) Alluvial
(c) Laterite (d) Black cotton

35. Laterite soil is majorly occurred in which state of India

(a) Kerala (b) Tamil Nadu
(c) Odisha (d) West Bengal

36. Colour denoted to the capability Class-1

(a) Light green (b) Yellow

(c) Red (d) Blue

37. Colour denoted to the Capability class-IV?

(a) Yellow (b) Red

(c) Blue (d) Green

38. Smallest unit of soil classification is?

(a) Soil series (b) Soil group

(c) Family (d) Order

39. Highest category of soil classification?

(a) Soil series (b) Soil group

(c) Family (d) Order

40. Humus contains?

(a) Lignin (b) Protein

(c) Both (d) Minerals

41. Which of the following nutrient available first to the plant?

(a) N (b) K

(c) P (d) all at same time

42. Which is correct for Saline soil

(a) pH>8.5, EC>4, ESP>15 (b) pH<8.5, EC>4, ESP<15

(c) pH>8.5, EC>4, ESP<15 (d) None of the above

43. Which of the following has highest anion exchange order?

(a) H_2PO_4 (b) Ca

(c) H (d) Mg

44. Alkali soils developed due to excess

(a) Na_2CO_3 (b) $NaHCO_3$

(c) Both (d) Iron pyrite

45. Salt sensitive fodder crop is

(a) Rhodes grass (b) Khush grass

(c) Sudan grass (d) Cluster bean

46. Electric Bike Taxi Scheme' is the initiative of which Indian state/UT?

(a) Tamil Nadu (b) Odisha

(c) Telangana (d) Karnataka

47. India's first Mobile Honey Processing Van was launched in which state?

(a) Uttar Pradesh (b) Gujarat

(c) Karnataka (d) Kerala

48. Which state declared four 'Vantangiya' villages as 'Revenue villages'?

(a) Gujarat (b) Uttar Pradesh

(c) Punjab (d) Assam

49. 'BA.1, BA.2 and BA.3' are the sub-lineages of which Covid Variant?

(a) Alpha (b) Gamma

(c) Delta (d) Omicron

50. As of January 2022, what is the minimum number of airbags compulsory for vehicles carrying up to eight passengers?

(a) One (b) Two

(c) Four (d) Six

ANSWER KEY

1	2	3	4	5	6	7	8	9	10	11	12	13	14	15
(c)	(b)	(a)	(b)	(a)	(a)	(a)	(b)	(a)	(b)	(d)	(d)	(d)	(a)	(a)
16	17	18	19	20	21	22	23	24	25	26	27	28	29	30
(b)	(c)	(a)	(a)	(d)	(a)	(b)	(c)	(b)	(b)	(a)	(a)	(a)	(d)	(a)
31	32	33	34	35	36	37	38	39	40	41	42	43	44	45
(a)	(b)	(a)	(c)	(a)	(a)	(c)	(a)	(d)	(c)	(a)	(b)	(a)	(c)	(d)
46	47	48	49	50										
(d)	(a)	(b)	(d)	(b)										

Chapter 9

Physical Classification of Water, Water Erosion, Mineralization, Immobilization, Aminization, Nitrification, Manures and Fertilizers, Green Manures, Fertilizer Classification, Fertilizer Grade and Fertilizer Ratio, Equivalent Acidity and Equivalent Basicity

1. **What are the targets in India renewable energy generation by 2030?**
   (a) 245 GW
   (b) 400GW
   (c) 450GW
   (d) 275GW
2. **Which organization has launched "Sankalp se sidhi village connect drive"?**
   (a) TRIFED
   (b) NAFED
   (c) NABARD
   (d) IFFCO
3. **The world food prize authority is situated at**
   (a) AUSTRALIA
   (b) KENIA
   (c) DES-MOINES
   (d) MEXICO
4. **Which of the following Indian received world food prize in the field of soil science recently?**
   (a) Shakuntala
   (b) Ratan lal
   (c) Groot
   (d) Nabarro and Haddad

5. Only Indian food brand to be globally recognized as both "clean label and 100 per cent whole grain"?

(a) True elements (b) Happilo

(c) Snakible (d) Whole truth

6. The transported silt soil due to wind is known as?

(a) Loess (b) Aolian

(c) Glacial (d) Alluvial

7. The accumulation of transported soil due to gravity is known as?

(a) Colluvial (b) Alluvial

(c) Glacial (d) Aeolian

8. The accumulation of transported soil due to Water is known as?

(a) Colluvial (b) Alluvial

(c) Glacial (d) Loess

9. Hygroscopic water held between tension of

(a) 31-10000 atm (b) 0.3 to 31 atm

(c) <3 atm (d) None of the above

10. Capillary water held at a tension of?

(a) 31-10000 atm (b) 0.3-31 atm

(c) <3 atm (d) None of the above

11. Which is the first stage of water erosion?

(a) Saltation (b) Splash erosion

(c) Sheet erosion (d) Rill erosion

12. Most serious form of water erosion for farmer?

(a) Splash (b) Rill

(c) Sheet (d) Gully

13. Which is the 1st stage of wind erosion?

(a) Splash erosion (b) Saltation

(c) Surface creep (d) Suspension

14. Permissible limit of soil loss in India (in ton/ha/yr)

(a) 10 (b) 12

(c) 14 (d) 16

15. Mineralization occurs when the C:N ratio is

(a) >30:1
(b) >20:1
(c) <20:1
(d) None of the above

16. Immobilization occurs when C:N ratio is

(a) >30:1
(b) <30:1
(c) <20:1
(d) None of the above

17. Maximum soil erosion due to which stage of wind erosion

(a) Saltation
(b) Surface creep
(c) Suspension
(d) None of the above

18. Which of the fertilizer is known as Kishan khad?

(a) CAN
(b) Ammonium nitrate
(c) Ammonium sulphate
(d) Sodium nitrate

19. Which of the following is explosive fertilizer?

(a) CAN
(b) Ammonium nitrate
(c) Ammonium sulphate
(d) Sodium nitrate

20. Equivalent acidity of Anhydrous ammonia is? (In meq/100g)

(a) 128
(b) 110
(c) 148
(d) 84

21. Equivalent acidity of urea?

(a) 128
(b) 110
(c) 84
(d) 77

22. Among the following fertilizer highest basic residual nature is seen in?

(a) Calcium cyanamide
(b) Sodium nitrate
(c) dicalcium phosphate
(d) calcium nitrate

23. Biurate percentage in urea should not exceed? (In percentage)

(a) 3
(b) 2
(c) 1.5
(d) 0.5

24. Potassium nitrate contains how much percentage of K_2O

(a) 13
(b) 44
(c) 22
(d) 25

25. Maximum consumption of N and P in the state?

(a) Uttar Pradesh
(b) Madhya Pradesh
(c) Maharashtra
(d) Andhra Pradesh

26. Minimum guarantee of plant nutrient content in term of available N, P and K is?

(a) Fertilizer grade
(b) Fertilizer ratio
(c) fertilizer reidue
(d) fertilizer order

27. Relative percentage of N, P and K is known as?

(a) Fertilizer grade
(b) Fertilizer Ratio
(c) Fertilizer Residue
(d) Fertilizer order

28. High analysis fertilizer contain more than how much percentage of primary fertilizer?

(a) 25
(b) 10
(c) 15
(d) 20

29. The ratio of N, P, K in FYM

(a) 0.5,0.25,0.5
(b) 7.3,1.5,1.3
(c) 3.0,1.0,1.5
(d) 5.5,4.0,2.0

30. *Sesbania rostrata* is which type of nodulating crop

(a) Stem nodulating
(b) Root nodulating
(c) Leaf nodulating
(d) None of the above

31. Example of Ex situ or green leaf manuring crop is?

(a) Sun hemp
(b) Dhaincha
(c) Cluster beans
(d) Glyricidia

32. Example of in situ green manuring crop

(a) Sunhemp
(b) Dhaincha
(c) Cluster beans
(d) All of the above

33. Optimum temperature for nitrifying bacteria is? (In degree C)

(a) 20-25
(b) 30-35
(c) 40-45
(d) 50-55

34. Optimum pH for nitrifying bacteria is?

(a) 4-4.5
(b) 5-5.5
(c) 6-6.5
(d) 6.5-7.5

35. Immobilization occurs when the C:N:P ratio exceeds

(a) 10:1:1
(b) 100:10:1
(c) 20:10:1
(d) 20:1:1

36. Percentage of phosphorus present in organic form in soil?

(a) 10-15 (b) 5-10

(c) 20-25 (d) 25-80

37. Conversion of ammonia to nitrite occurs by the microbe?

(a) Nitrosomonas (b) Nitrobacter

(c) Both (d) None of the above

38. Conversion of nitrite to nitrate occurs by the microbe?

(a) Nitrosomonas (b) Nitrobacter

(c) Both (d) None of the above

39. Reverse of mineralization where available form of an element is fixed as immobile form?

(a) Mineralization (b) Immobilization

(c) Extra-mineralization (d) None of the above

40. When erosion is caused by excessive grazing, deforestation *etc.* is called

(a) Artificial erosion (b) Natural eroion

(c) Anthropogenic erosion (d) None of the above

41. For foliar spray of urea, Biurate concentration should not exceed (in percentage)

(a) 1.5 (b) 0.5

(c) 0.25 (d) 1

42. Leaching losses are maximum for which kind of fertilizer?

(a) Nitrate (b) Nitrite

(c) Ammonical (d) Aminical

43. Sodium nitrate contains how much percentage of N?

(a) 8 (b) 16

(c) 24 (d) 30

44. Which fertilizer is suitable for water logged soil?

(a) Ammonical (b) Nitrate

(c) Nitrite (d) Amminical

45. Equivalent basicity of calcium nitrate?

(a) 29 (b) 25

(c) 21 (d) 16

46. Prabha Atre, who was conferred the Padma Vibhushan is associated with which field?

(a) Literature (b) Music

(c) Sports (d) Civil Service

47. Scientists from IIT Kharagpur have developed a portable device for detecting which disease?

(a) Oral Cancer (b) Breast Cancer

(c) Prostate Cancer (d) Lung Cancer

48. As per IMF World Economic Outlook January assessment, what is the estimated growth of India in 2021-22?

(a) 8.5 per cent (b) 9 per cent

(c) 9.5 per cent (d) 10 per cent

49. Which country is associated with the "Aboriginal flag", which was seen in the news recently?

(a) USA (b) UK

(c) Australia (d) China

50. *Halodule uninervis*, a species of sea grass, is found to have strong activity against which disease?

(a) Cancer (b) Hypertension

(c) Diabetes (d) Covid 19

ANSWER KEY

1	2	3	4	5	6	7	8	9	10	11	12	13	14	15
(c)	(a)	(c)	(b)	(a)	(a)	(a)	(b)	(a)	(b)	(b)	(c)	(b)	(b)	(c)
16	17	18	19	20	21	22	23	24	25	26	27	28	29	30
(a)	(a)	(a)	(b)	(c)	(c)	(a)	(c)	(b)	(a)	(a)	(b)	(a)	(a)	(a)
31	32	33	34	35	36	37	38	39	40	41	42	43	44	45
(d)	(d)	(b)	(d)	(b)	(d)	(a)	(b)	(b)	(c)	(c)	(a)	(b)	(a)	(c)
46	47	48	49	50										
(b)	(a)	(b)	(c)	(a)										

Chapter 10

Nitrogen Losses, Nitrification Inhibitors, Slow-Release Fertilizer, Phosphate Fertilizer, Activity Ratio, Methods of Fertilizer Application, Biofertilizers, Soil Microorganisms, Classifications of Plant Nutrients, Law of Minimum, Criteria for Essentiality, Role of Nutrient and Deficiency Symptoms, Terminologies

1. **Sulphur, Rock phosphate and Thiobacillus were combined to make this product**
   (a) Pelophos (b) DCP
   (c) Biosuper (d) None of the above
2. **Which state has the highest use of potassic fertilizer?**
   (a) West Bengal (b) Punjab
   (c) Maharashtra (d) Telangana
3. **A slow-release N fertilizer that works by the principle of nitrification inhibition**
   (a) Urea formaldehyde (b) N-serve
   (c) PMA (d) Sulphur coated urea
4. **Which among the following is not a nitrification inhibitor?**
   (a) N-serve (b) Pyrimidine
   (c) Dicyandiamide (d) PPDA (Phenyl phosphoro-diamidate)

5. Potassic fertilizer most suitable for tomato and potato

(a) Muriate of potash (b) Schoenite

(c) Sulphate of potash (d) All of the above

6. P2O5 content of Rock phosphate

(a) 10 to 20 per cent (b) 20 to 30 per cent

(c) 20 to 40 per cent (d) 30 to 50 per cent

7. The conversion factor for calculating P from P_2O_5

(a) P_2O_5 * 2.29 (b) P_2O_5 * 0.43

(c) P * 2.29 (d) P * 0.43

8. Percent loss of nitrogen in India due to volatilization

(a) 30 per cent (b) 40 per cent

(c) 50 per cent (d) 60 per cent

9. Analytical method for determining available Phosphorus in alkaline soil

(a) Olsen's method (b) Bray and Krutz method

(c) Mehlich's method (d) All of the above

10. Phosphatic fertilizer that also provides sulphur is

(a) Rock phosphate (b) Single super phosphate

(c) DCP (d) DAP

11. Gypsum ($CaSO_4 \cdot 2H_2O$) used for reclamation purpose in

(a) Acidic soil @ 800 kg/ha (b) Alkali soil @ 250 kg/ha

(c) Alkali soil @ 800 kg/ha (d) Acidic soil @ 250 kg/ha

12. Percentage of boron in borax

(a) 10 (b) 11

(c) 12 (d) 13

13. The term functional nutrient was given by

(a) Arnon and Stout (b) Sprengel

(c) Justus Von Liebig (d) Nicholas

14. The nutrient which is absorbed as neutral molecule

(a) Boron (b) Iron

(c) Potassium (d) Calcium

15. Plant absorbs phosphorus in the form of

(a) HPO ^{2}{
(b) $PO4^{3-}$
(c) H_2PO^{4-}
(d) Both (a) and (c)

16. Indicator plant for phosphorus deficiency

(a) Mustard
(b) Cauliflower
(c) Potato
(d) Sunflower

17. Which among the following is micro-aerobic nitrogen fixing microbe?

(a) Rhizobium
(b) Azospirillum
(c) Anabaena
(d) Azotobacter

18. The percentage of plant nutrients in a fertilizer is called

(a) Fertilizer grade
(b) Fertilizer ratio
(c) Agriculture grade
(d) Activity ratio

19. The second most deficient nutrient in Indian soils

(a) Iron
(b) Nitrogen
(c) Zinc
(d) Copper

20. The relative proportion of NPK that has to be applied as starter dose in legumes and vegetable seedling at transplanting is

(a) 1: 1: 1
(b) 1: 2: 1
(c) 2: 1: 1
(d) 2: 2: 1

21. A free living anaerobic bacteria capable of nitrogen fixation

(a) Rhizobium
(b) Clostridium
(c) Streptococcus
(d) Azotobacter

22. Common biostrain of Rhizobium that nodulates groundnut is

(a) *R. leguminosarum*
(b) *R. meliloti*
(c) *R. phaseoli*
(d) *R. japonicum*

23. Recently recognized 17th essential element is

(a) Cl
(b) Na
(c) Ni
(d) B

24. Nitrogen percentage in thiourea is

(a) 32.2 per cent
(b) 32.5 per cent
(c) 36.8 per cent
(d) 31.0 per cent

25. Potassic fertilizers are not suitable for

(a) Acidic soils
(b) Heavy soils
(c) Both (a) and (b)
(d) Alkaline soils

26. Indicator plant for potassium

(a) Mustard
(b) Cauliflower
(c) Potato
(d) Sunflower

27. When elemental sulphur is applied to soil, _______ converts it to sulphate

(a) Thiobacillus
(b) Desulfovibrio
(c) Bacillus
(d) Sulfobacillus

28. Diethylene triamine Penta acetic acid (DTPA) is used for the estimation of available

(a) Fe, Mn, Cu, Zn, B, Mo and Cl
(b) Fe, Mn, Cu, Zn
(c) N, P, K
(d) Ca, Mg, S

29. The most abundant organism in soil

(a) Bacteria
(b) Fungi
(c) Actinomycetes
(d) Algae

30. In nitrification process, conversion of nitrite to nitrate is done by

(a) Nitrosomonas
(b) Nitrobacter
(c) Pseudomonas
(d) Bacillus

31. Which among the following is a denitrifying bacterium?

(a) Pseudomonas
(b) Paracoccus
(c) Bacillus
(d) All of the above

32. Which crop is capable of utilizing ammonia effectively?

(a) Paddy
(b) Wheat
(c) Jowar
(d) Tea

33. Most of the plants absorb nutrients in

(a) Reduced form
(b) Oxidized form
(c) Both (a) and (b)
(d) None of the above

34. Ammonia volatilization is favoured by

(a) Alkaline condition
(b) Acidic condition
(c) Neutral condition
(d) None of the above

35. Which among the following is a citrate soluble P fertilizer?

(a) Rock phosphate
(b) Basic slag
(c) Mono ammonium phosphate
(d) Raw Bone meal

36. Rate at which gypsum is applied to crops, especially onion, mustard, groundnut and other pulse crops

(a) 200 kg/ha
(b) 250 kg/ha
(c) 300 kg/ha
(d) 350 kg/ha

37. In anaerobic soils, sulphates are converted to H_2S by the action of

(a) Bacillus
(b) Thiobacillus
(c) Desulfovibrio
(d) Sulfobacillus

38. *Frankia* is

(a) Actinomycete
(b) Cyanobacterium
(c) Symbiotic bacteria
(d) Non symbiotic bacteria

39. Who first isolated nitrifying bacteria responsible for production of nitrate from ammonia?

(a) Carnaham
(b) Winogradsky
(c) Del Tredici
(d) Doberiner

40. The idea that plants need essential minerals was first proposed by

(a) Aristotle
(b) Bernand Simpson
(c) Arnon and Stout
(d) Von Haier

41. Country's first grain ATM installed in which city?

(a) Bengaluru, Karnataka
(b) Jaipur, Rajasthan
(c) Kevadia, Gujarat
(d) Gurugram, Haryana

42. Tequila fish, which was declared extinct has been reintroduced to which country?

(a) Argentina
(b) Mexico
(c) Chile
(d) Russia

43. India's first pollinator park has been established in the city of

(a) Jaipur, Rajasthan
(b) Karnal, Haryana
(c) Nainital, Uttarakhand
(d) Guwahati, Assam

44. Geological survey of India has approved to set up India's first Geo- Park in which state?

(a) Maharashtra
(b) Goa
(c) Madhya Pradesh
(d) Telangana

45. What is the name of the Tropical Storm which made landfall on three southern African countries?

(a) Rina (b) Ana

(c) Boss (d) Sue

46. Which online facility has been launched by the Union Government to provide a direct solution to the problem of agriculture sector?

(a) e-Krishi Samasya (b) e-Krishi Samvad

(c) e-Krishi Samveda (d) e-Krishi Solution

47. International Year of Millets

(a) 2020 (b) 2021

(c) 2022 (d) 2023

48. First state in the country to flag of India's first fruit train

(a) Andhra Pradesh (b) Tamil Nadu

(c) Karnataka (d) Maharashtra

49. Which of the following national parks is not a UNESCO World Heritage site?

(a) Kaziranga (b) Keoladeo

(c) Sundarbans (d) Kanha

50. Which year did the tragic Bhopal gas disaster occur?

(a) 1976 (b) 1984

(c) 1991 (d) 1981

ANSWER KEY

1	2	3	4	5	6	7	8	9	10	11	12	13	14	15
(c)	(c)	(b)	(d)	(c)	(c)	(b)	(d)	(a)	(b)	(c)	(b)	(d)	(a)	(d)
16	17	18	19	20	21	22	23	24	25	26	27	28	29	30
(a)	(b)	(a)	(c)	(b)	(b)	(d)	(c)	(c)	(d)	(c)	(a)	(b)	(a)	(b)
31	32	33	34	35	36	37	38	39	40	41	42	43	44	45
(d)	(a)	(a)	(a)	(b)	(b)	(d)	(a)	(b)	(c)	(d)	(b)	(c)	(c)	(b)
46	47	48	49	50										
(b)	(d)	(a)	(d)	(b)										

PLANT BREEDING AND GENETICS

Chapter 11

Important Points of Genetics, Organs and Discovered by, Premendelian Hereditary Theories, Cell Wall, Plasma Lemma, Cytoplasm, Endoplasmic Reticulum, Ribosomes, Golgibody, Lysosome, Peroxisome, Glyoxysome, Vacuole, Mitochondria, Chloroplast

1. **Actual site of photosynthesis is**
   (a) Mitochondria
   (b) Chloroplast
   (c) Golgi body
   (d) Lysosomes
2. **Tonoplast is the membrane surrounding**
   (a) Peroxisome
   (b) Spherosomes
   (c) Vacoule
   (d) Centriole
3. **Endomembrane system of cells is**
   (a) Vacoule
   (b) Centriole
   (c) Endoplasmic reticulum
   (d) Golgi body
4. **Power house of cell is**
   (a) Chloroplasts
   (b) Mitochondria
   (c) Cytoplasm
   (d) Endoplasmic reticulum
5. **Following are found in animal cells only**
   (a) Endoplasmic reticulum
   (b) Lysosomes
   (c) Glyoxisomes
   (d) Centriole

6. **Autolysis of cell is a function of**
   (a) Peroxisome (b) Mitochondria
   (c) Lysosomes (d) Ribosomes

7. **Largest cell is**
   (a) Ostrich egg, 20 mm (b) Ostrich egg, 35 mm
   (c) Ostrich egg, 75 mm (d) Ostrich egg, 90 mm

8. **Following organelle is lacking in prokaryote**
   (a) ER (b) Mitochondria
   (c) Lysosomes (d) All of the above

9. **what is synthesized by smooth ER (Endoplasmic Reticulum)**
   (a) Cholestrol (b) Testosterone
   (c) Proteins (d) Both (a) and (b)

10. **ER is absent in**
    (a) Red blood cells (b) White blood cells
    (c) Monocytes (d) All of the above

11. **Sub units of 80 S ribosomes are**
    (a) 50 S and 30 S (b) 40 S and 40 S
    (c) 60 S and 40 S (d) 50 S and 40 S

12. **Mitochondrial ribosome contain sub units**
    (a) 60 S and 40 S (b) 50 S and 30 S
    (c) 40 S and 40 S (d) 50 S and 40 S

13. **Non-living part of a cell is**
    (a) Cell membrane (b) Tonoplast
    (c) Nuclear membrane (d) Cell wAll of the above

14. **Middle lamella contains**
    (a) Ca (b) Mg
    (c) Both (a) and (b) (d) Na

15. **Lipid and protein content of plasma membrane are in the ratio of**
    (a) 1:2 (b) 3:2
    (c) 2:3 (d) 2:1

16. **Lysosomes are synthesized from**
    (a) Peroxisome (b) ER
    (c) Mitochondria (d) Golgi complex

17. Fats are stored in

(a) Chromoplasts
(b) Mitochondria
(c) SER
(d) Leucoplasts

18. Phycocyanin is present in

(a) Vacoule
(b) Chromoplasts
(c) Leucoplasts
(d) All of the above

19. Inner most layer of cell wall is

(a) Middle lamella
(b) Primary cell wall
(c) Secondary cell wall
(d) None of the above

20. The following acts as a transport vehicle across organelles

(a) Middle lamella
(b) Primary cell wall
(c) Secondary cell wall
(d) Plasma lemma

21. Main enzymes present in lysosomes are

(a) Alkaline phosphatase
(b) Acid phosphatase
(c) Phosphorous acid
(d) All of the above

22. Cell wall formation is supported by

(a) Golgi body
(b) ER
(c) Mitochondria
(d) Vacoule

23. Following is correct combination

(a) Spherosomes - only in plant cells, centrioles - only in animal cells
(b) Spherosomes - only in animal cells, centrioles - only in plant cells
(c) Both organelles are present only in plant cells
(d) Both organelles are present only in animal cells

24. Organelle concerned with photorespiration is

(a) Peroxisome in C_4 plants
(b) Peroxisome in C_3 plants
(c) Vacoule in C_3 plant
(d) Vacoule in C_4 plant

25. Crystae is

(a) Inner folds of inner membrane of mitochondria
(b) Inner folds of outer membrane of mitochondria
(c) Outer folds of inner membrane of mitochondria
(d) Outer folds of outer membrane of mitochondria

26. Oxysomes are associated with

(a) Electron transport particles
(b) Oxygen transport particles
(c) Proton transport particles
(d) None of the above

27. Respiratory complex contains

(a) NADP dehydrogenase
(b) b-c 1 complex
(c) Cytochrome reductase complex
(d) All of the above

28. Glycolate metabolism is regulated by

(a) Glyoxysomes
(b) Spherosomes
(c) Vacoules
(d) None of the above

29. Number of mitochondria in a cell

(a) 1
(b) 50-100
(c) 200-800
(d) 1000-1800

30. Site of krebs cycle is

(a) Mitochondria
(b) Plasmalemma
(c) Chloroplast
(d) All of the above

31. Microtubules are absent in cytoplasm of

(a) Prokaryotes
(b) Eukaryotes
(c) Both (a) and (b)
(d) None of the above

32. Father of genetics is

(a) Mendel
(b) Haber land
(c) Theophrastus
(d) Aristotle

33. Genetic composition of bacteriophages is

(a) ds DNA
(b) ss DNA
(c) ss RNA
(d) ds RNA

34. Moat abundant form of RNA is

(a) t RNA
(b) m RNA
(c) r RNA
(d) All of the above

35. Protoplasm is

(a) Cytoplasm + mitochondria
(b) Cytoplasm + nucleus
(c) Cytoplasm + ER
(d) Cytoplasm only

36. Following is correct combination

(a) Nucleus-Robert brown, cell wall- Robert hook

(b) Cell wall-Robert brown, nucleus- Robert hook

(c) Both cell wall and nucleus- Robert brown

(d) Both cell wall and nucleus- Robert hook

37. Fluid mosaic model was given by

(a) Singer (b) Nicholson

(c) Porter (d) Both (a) and (b)

38. Chromosome term given by

(a) Waldeyer (b) Strasburger

(c) Both (a) and (b) (d) Boveri

39. Cellulose content in primary cell wall is

(a) 100 per cent (b) 20 per cent

(c) 30 per cent (d) 50 per cent

40. Number of photosynthetic units in chloroplasts is

(a) 100-150 (b) 150-200

(c) 250-300 (d) 450-500

41. NAFED stands for

(a) National Agricultural Federation

(b) National Agricultural Marketing Federation

(c) National Agricultural Cooperative Federation

(d) National Agricultural Marketing Cooperative Federation

42. Crop with highest pesticidal application

(a) Cotton (b) Paddy

(c) Wheat (d) Maize

43. Total number of CIPMCs at present is

(a) 30 (b) 25

(c) 35 (d) 40

44. First cereal to be cultivated by man

(a) Wheat (b) Rice

(c) Maize (d) None of the above

45. First tree city of India

(a) Pune
(b) Delhi
(c) Hyderabad
(d) Goa

46. First bio aesthetically planned city

(a) Hyderabad
(b) Goa
(c) Surat
(d) Chandigarh

47. First crop grown in India

(a) Wheat
(b) Rice
(c) Maize
(d) Sugarcane

48. First completely organic country is

(a) India
(b) Bhutan
(c) China
(d) Nepal

49. First organic state of India

(a) Sikkim
(b) Goa
(c) Uttar Pradesh
(d) Andhra Pradesh

50. First fertilizer manufactured in India

(a) Urea
(b) SSP
(c) DAP
(d) MOP

ANSWER KEY

1	2	3	4	5	6	7	8	9	10	11	12	13	14	15
(b)	(c)	(d)	(b)	(d)	(c)	(c)	(d)	(d)	(a)	(c)	(b)	(d)	(c)	(b)
16	17	18	19	20	21	22	23	24	25	26	27	28	29	30
(d)	(d)	(b)	(c)	(d)	(b)	(a)	(a)	(b)	(a)	(a)	(b)	(d)	(c)	(a)
31	32	33	34	35	36	37	38	39	40	41	42	43	44	45
(a)	(a)	(b)	(c)	(b)	(a)	(d)	(a)	(c)	(c)	(d)	(b)	(c)	(a)	(c)
46	47	48	49	50										
(d)	(b)	(b)	(a)	(b)										

Chapter 12

Cell Division, Cell Cycle, Mitosis, Meiosis, Chromosome Structure and Function, Chromosomal Aberration, DNA, RNA, Deletion, Duplication, Inversion, Translocation

1. The term mitosis was first used by which of the following

(a) Mendel
(b) Flemming
(c) Strasburger
(d) Boveri

2. How much time does G1 phase occurs in the cell replication

(a) 6-12 hours
(b) 1-2 hours
(c) 3-4 hours
(d) 12-24 hours

3. Which of the following scientist reported the chromosome 21 is involved in down's syndrome

(a) C.B. Bridges
(b) L. Down
(c) J. Lejeune and Co-workers
(d) C. Stem

4. Which of the following is the longest stage of cell cycle

(a) G1 phase
(b) Prophase
(c) Meta phase
(d) Anaphase

5. Which is the shortest stage of cell cycle

(a) Prophase
(b) Metaphase
(c) Anaphase
(d) Teleophase

6. **Which of the following are the characters of mitosis**
   (a) Chromosome number remain same
   (b) DNA remains identically same
   (c) One cell gives rise to two cells
   (d) All of the above options are correct

7. **A typical cell goes through a process of growth, development and reproduction called as**
   (a) Mitosis (b) Meiosis
   (c) Cell cycle (d) Interphase

8. **which of the following produces the identical twins**
   (a) Mitosis (b) Meiosis
   (c) Both (d) None of the above

9. **Condensation of chromosome, disappearance of nucleolus are the main features seen in**
   (a) Prophase (b) Meta phase
   (c) Annaphase (d) Teleophase

10. **Formation of cleavage and separation of cell is mainly seen in which type of cell**
    (a) Animal cell (b) Bacterial cell
    (c) Protoplast cell (d) Viral cell

11. **Meiosis is generally occurring in which cells**
    (a) Vegetative cells
    (b) Reproductive cell
    (c) In both vegetative and reproductive cells
    (d) None of the above

12. **If a cell is having a basic chromosome number of 20. If cell undergoes mitosis the what will be the number of chromosomes after the cell division**
    (a) 20 (b) 40
    (c) 60 (d) 80

13. **Meiosis was first described in which of the following**
    (a) Maize (b) Sea urchin eggs
    (c) Sea horse eggs (d) None of the above

14. Meiosis was first discovered in 1876 by

(a) William Heswing (b) Oscar Hertwing

(c) Mendel (d) J.B. Moore

15. The term meiosis was coined by

(a) J.B. Farmer (b) J.B. Moore

(c) Oscar Hertwing (d) Both (a) and (b)

16. Individual condensation of chromosome into long strand with in the nucleus occurs in which stage of meiosis

(a) Leptotene (b) Zygotene

(c) Pachytene (d) Diakinensis

17. In pachytene the homologous chromosome become much more closely associated this process is called as

(a) Diapsis (b) Synapses

(c) Diakinensis (d) Zyygotene

18. The interchange of part of chromatid of a pair of homologous chromosomes is called as

(a) Synapses (b) Crossing over

(c) Zyotene (d) Diakinensis

19. Chromosome was first discovered by which of the following

(a) Walderyer (b) Strasburger

(c) Bateson (d) Bovery

20. During cell division spindle fibre is attached to which of the following

(a) Centromere (b) Teleomerre

(c) Choromere (d) Nuclesome

21. Which is not correct about centromere

(a) Centromere is located in center of chromosome

(b) During anaphase the chromosome appears j-shape

(c) Both the arms of the chromosome are same

(d) Arm ratio is one

22. Poly centric chromosomes are found in which of the following

(a) Thyanata (b) Luzula

(c) Generative tissue of parascris equorum (d) All of the above

23. Which of the following animals contain the lowest chromosome number in the somatic tissue

(a) Drosophila (b) Housefly

(c) Round worm (d) Toad

24. Which of the following plants contain highest chromosome number in their somatic tissue

(a) Piper nigrum (b) Strawberry

(c) Adder's tongue fern (d) Sugarcane

25. Which of the following below options are correct about importance of crossing over

(a) Produces a new individual having a new combination of traits

(b) Helps in chromosome mapping

(c) Both (a) and (b) are correct

(d) None of the above

26. Crossing over occurs in which division of meiosis

(a) 1 (b) 2

(c) 3 (d) 4

27. Nucleolus organiser region (NOR) of sat chromosome contains gene for the synthesis of which of the following

(a) rRNA (b) tRNA

(c) mRNA (d) SnRNA

28. The folded fibre model of chromosome structure was proposed by which of the following

(a) Dupraw (b) Kornberg and Thomas

(c) Melson and Stahi (d) J. H. Taylor

29. Which of the following are distinct from rest

(a) Chromomere (b) Centromere

(c) Kinetochore (d) Primary constriction

30. Who postulated that the loss of telomere makes chromosome end unstable

(a) Muller (b) Morgan

(c) Stadler (d) Stern

31. In Patau's syndrome which of the following chromosome is involved

(a) 21 (b) 13

(c) 18 (d) 16

32. Philadelphia 22and cry -da-chat syndrome of humans are examples of which of the following

(a) Duplication (b) Translocation

(c) Inversion (d) Deficiency

33. Turner's syndrome is associated with each of the following

(a) An autosome (b) X- chromosome

(c) Y- chromosome (d) Both x and y chromosome

34. Haploids can be used for the production of which of the following

(a) Inbreeds (b) Homozygous lines

(c) Hybrids (d) Synthetics

35. Facultative heterochromatin is associated with which of the following

(a) Dosage competition (b) Position effect

(c) Transposition (d) Both (a) and (c)

36. Which of the following is presumably an autotetraploid

(a) Lucerne and peanut (b) Potato and Peanut

(c) Potato and Coffee (d) All of the above

37. In which of the following a complete set of monosomics is available

(a) Wheat and oats (b) Tobacco

(c) Cotton (d) All of the above

38. In which of the following DNA form unusual structure

(a) Centomere (b) Knob

(c) Chromere (d) Telomere

39. Nullisomic have been developed in which of the following

(a) Wheat (b) Cotton

(c) Oats (d) Both (a) and (c)

40. Nullsomics are represented by which of the following

(a) 2n+2 (b) 2n-2

(c) 2n-1+1 (d) 2n-1-1

41. How many monosomics can be developed in bread wheat

(a) 21 (b) 41

(c) 42 (d) 20

42. Which of the following histone is the most easily separated from chromatin

(a) H2b (b) H1

(c) H3 (d) H4

43. A triple monosomic can be represented by.

(a) 2n-2-1 (b) 2n-1-1-1

(c) 2n-1-2 (d) 2n-3

44. During 1923 translocation was discovered by which of the following

(a) Belling (b) Bridges

(c) Sturtevant (d) Morgan

45. Which of the following produces the pseudo dominance

(a) Inversion (b) Deletion

(c) Translocation (d) Duplication

46. Inversion loops are observable during which of the following stages of meiosis

(a) Diplotene (b) Pachytene

(c) Leptotene (d) Diakinesis

47. Inversion leads to change in

(a) Number of genes in chromosome (b) Kind of gene in chromosome

(c) Order of gene in chromosome (d) Chromosome length

48. Which of the following can change the chromosome structure

(a) Translocation (b) Inversion

(c) Duplication (d) Deletion

49. which of the following is essential for normal chromosome behaviour and/or stability

(a) Telomere (b) Chromomere

(c) Centromere (d) Both (b) and (c)

50. Which of the following is not found in normal chromosome

(a) Secondary constriction (b) Chromid

(c) Telomere (d) Centromere

ANSWER KEY

1	2	3	4	5	6	7	8	9	10	11	12	13	14	15
(b)	(a)	(c)	(a)	(c)	(d)	(c)	(a)	(a)	(a)	(a)	(b)	(b)	(b)	(d)
16	17	18	19	20	21	22	23	24	25	26	27	28	29	30
(a)	(b)	(b)	(a)	(a)	(b)	(d)	(c)	(c)	(c)	(a)	(a)	(a)	(a)	(a)
31	32	33	34	35	36	37	38	39	40	41	42	43	44	45
(b)	(d)	(d)	(b)	(d)	(d)	(d)	(d)	(d)	(b)	(a)	(b)	(b)	(b)	(b)
46	47	48	49	50										
(b)	(c)	(d)	(d)	(a)										

Chapter 13

Genome, Polyploidy, Protein Synthesis, Transcription, Replication, Translation, Classical Genetics, Mendellian Laws, Multiple Alleles, Pseudo Allels, Type of Interaction

1. **Autotriploids are generally**
   (a) Highly sterile
   (b) Highly fertile
   (c) Homologous chromosome
   (d) Both (a) and (c)

2. **The example of Autotriploids**
   (a) Banana
   (b) Sugarcane
   (c) Apple
   (d) All of the above

3. **Mostly cultivated varieties of banana are**
   (a) Autotriploids
   (b) Allotriploids
   (c) Autotetraploids
   (d) Allotetraploids

4. **Examples of Autotetraploids**
   (a) Rye
   (b) Grasses
   (c) Alfa alfa
   (d) All of the above

5. **The example of allotetraploids**
   (a) Tobacco
   (b) Brassica
   (c) Cotton
   (d) All of the above

6. **Triticale is example of**
   (a) Inter generic polyploidy
   (b) Inter specific polyploidy
   (c) Intra generic polyploidy
   (d) Intra specific polyploidy

7. Triticale is cross between

(a) Wheat and rye (b) Wheat and rice
(c) Rice (d) All of the above

8. Monosomic condition is

(a) 2n-1 (b) 2n-1-1
(c) n+1 (d) None of the above

9. Examples of double monosomy is

(a) 2n-2 (b) 2n-1-1
(c) 2n+1 (d) 2n+2

10. The condition of nullisomic is

(a) 2n-1-1 (b) 2n-2
(c) 2n+2 (d) 2n+1+1

11. The condition of trisomic condition is

(a) 2n-1-1 (b) 2n+1
(c) 3n (d) 2n+1+1

12. Downs syndrome is due to which condition of chromosomes

(a) Trisomics (b) Nullisomic
(c) Tetrasomic (d) Monosomic

13. Back cross is

(a) F1 x dominant parents (b) F1 x recessive parents
(c) Both (d) None of the above

14. Test cross is

(a) F1 x dominant parents (b) F1 x recessive parents
(c) Both (d) None of the above

15. (A x B) x (C x D) is example of

(a) Three-way cross hybrid (b) Double cross hybrid
(c) Single cross hybrid (d) Modified single cross hybrid

16. (A x B) x C

(a) Three-way cross hybrid (b) Double cross hybrid
(c) Single cross hybrid (d) Modified single cross hybrid

17. A cross between inbred line and open pollinated variety is known as

(a) Back cross (b) Top cross
(c) Test cross (d) None of the above

18. Chromosome number of fruit fly is (2n)

(a) 8 (b) 32

(c) 4 (d) 16

19. Chromatin is composed of

(a) DNA (b) RNA

(c) Protein (d) All of the above

20. Histones constitute about how much percentage of chromosomal protein

(a) 60 (b) 70

(c) 80 (d) 95

21. Examples of multiple alleles is

(a) Fur colour in rabbit (b) Wing type drosophila

(c) Eye colour of drosophila (d) All of the above

22. Complementary gene action ratio is

(a) 9:3:4 (b) 9:7

(c) 13:3 (d) 9:6:1

23. Inhibitory gene action ratio is

(a) 9:7 (b) 9:3:4

(c) 13:3 (d) 9:6:1

24. Duplicate gene action ratio is

(a) 9:7 (b) 15:1

(c) 9:3:4 (d) 13:3

25. Masking gene action ratio is

(a) 12:3:1 (b) 9:7

(c) 15:1 (d) 9:3:4

26. Which law is also known as law of purity of gametes?

(a) Law of dominance (b) Law of segregation

(c) Law of independent assortment (d) None of the above

27. Cis trans position effect is seen in

(a) Pseudo alleles (b) Multiple alleles

(c) Incomplete dominance (d) None of the above

28. Both genes have independent dominant effect is proved by

(a) Law of independent assortment
(b) Law of dominance
(c) Law of segregation
(d) Law of tolerance

29. Smallest and functionally part of genetic material is

(a) Gene
(b) Genome
(c) Genomics
(d) All of the above

30. An alternative form of gene is known as

(a) Genome
(b) Genomics
(c) Allele
(d) None of the above

31. Polyploidy seen in common bread wheat is

(a) Allohexaploid
(b) Autohexoaploid
(c) Autotriploids
(d) Autotetraploids

32. A single set of chromosomes is

(a) Haploid
(b) Monoploid
(c) Double haploid
(d) Both (a) and (b)

33. Monoploid is

(a) Fertile
(b) Sterile
(c) Over fertile
(d) None of the above

34. Sugar beet contain large root and more sugar contain are

(a) Triploids
(b) Tetraploid
(c) Haploid
(d) None of the above

35. Mostly cultivated and seedless banana are

(a) Triploids
(b) Tetraploid
(c) Haploid
(d) All of the above

36. Early maturing, disease resistance and seedless watermelon are

(a) Triploids
(b) Tetraploid
(c) Hexaploid
(d) Diploid

37. Groundnut, potato and grasses are the example of

(a) Auto tetraploid
(b) Allotriploids
(c) Allotetraploids
(d) Hexaploid

38. The basic set of chromosomes of an individual is known as

(a) Genome
(b) Gene
(c) Alleles
(d) None of the above

39. Polyploidy is induced by the chemical

(a) Colchicine (b) Sulphonate
(c) Aluminium phosphate (d) None of the above

40. Increase in vigour, size in autopolyploid is known as

(a) Gigantism (b) Syndrome
(c) Hybrid vigour (d) None of the above

41. First Nobel prize by agriculturist

(a) N.E. Borlaug (b) M.S. Swaminathan
(c) B.P. Pal (d) Viswanathan

42. First world food prize recipient

(a) N.E Borloug (b) M.S. Swaminathan
(c) B.P. Pal (d) Viswanathan

43. First Agriculture minister of independent India is

(a) K.M. Manushi (b) Rajendra Prasad
(c) Jairamraj (d) Narendra Singh Tomar

44. First organic farming University is going to be established at

(a) Vadodara (b) Dantewada
(c) Hyderabad (d) Bhubaneshwar

45. First crop raised by hydroponics is

(a) Tomato (b) Brinjal
(c) Rice (d) Potato

46. In which Indian state the Arjuna Sahayak irrigation project is being implemented?

(a) Andhra Pradesh (b) Uttar Pradesh
(c) Haryana (d) Himachal Pradesh

47. What is India's target for ethanol blending in petrol for the year 2022?

(a) 1 per cent (b) 1.5 per cent
(c) 2 per cent (d) 10 per cent

48. Indian Army has inaugurated the First Green Solar Energy harnessing plant in which state?

(a) Rajasthan (b) Sikkim
(c) Punjab (d) Gujarat

49. Which body permits Telecommunication companies to conduct trials for 5G technology in India?

(a) Department of Science and Technology

(b) Department of Telecommunications

(c) TRAI

(d) NITI Aayog

50. KVK of which state has for the first time introduced Hydrogel capsules, for soil conservation?

(a) Tamil Nadu (b) Kerla

(c) Madhya Pradesh (d) Uttar Pradesh

ANSWER KEY

1	2	3	4	5	6	7	8	9	10	11	12	13	14	15
(d)	(d)	(a)	(d)	(d)	(a)	(a)	(a)	(b)	(b)	(b)	(a)	(c)	(b)	(b)
16	17	18	19	20	21	22	23	24	25	26	27	28	29	30
(a)	(b)	(a)	(d)	(c)	(d)	(b)	(c)	(b)	(a)	(b)	(a)	(a)	(a)	(c)
31	32	33	34	35	36	37	38	39	40	41	42	43	44	45
(a)	(d)	(b)	(a)	(a)	(a)	(a)	(a)	(a)	(a)	(a)	(b)	(b)	(a)	(a)
46	47	48	49	50										
(b)	(d)	(b)	(b)	(b)										

Chapter 14

Mutation, Sex Determination, Sex Linked and Sex-Limited Characters, Plant Breeding History, Breeding Techniques, Pureline, Mass Selection, Clonal Selection, Pedigree Method, Bulk Breeding, Backcross, Recurrent Selection

1. **Center of origin was given by**
   (a) Charles darwin
   (b) Vavilon
   (c) Both (a) and (b)
   (d) Strasburger

2. **Tenalism is**
   (a) mRNA to DNA
   (b) DNA to mRNA
   (c) mRNA to protein synthesis
   (d) None of the above

3. **Triticale is**
   (a) Interspecific hybrid
   (b) Intergeneric hybrid
   (c) Intraspecific hybrid
   (d) None of the above

4. **Sex determining genes are present on**
   (a) X chromosome
   (b) Autosomes
   (c) Both (a) and (b)
   (d) None of the above

5. **Mutation is**
   (a) Sudden, non-heritable
   (b) Cumulative, heritable
   (c) Sudden, heritable
   (d) Cumulative, non-heritable

6. Non ionising radiations are

(a) Alpha rays
(b) Beta rays
(c) X rays
(d) UV rays

7. Pick the odd one out with respect to categorization

(a) EMS
(b) MMS
(c) Mustard gas
(d) Ethidium bromide

8. A phenomenon opposite to crossing over is. which is detected by.

(a) Interference, test cross
(b) Linkage, back cross
(c) Interference, back cross
(d) Linkage, test cross

9. The rate of mutation in bacteria is

(a) 1 in 100
(b) 1 in 10000
(c) 1 in 10000000
(d) 1 in 100000000000

10. Following combination is correct regarding linkage phases

(a) Coupling: between two dominant alleles, repulsion: between dominant and recessive
(b) Coupling: between dominant and recessive, repulsion: between 2 dominant alleles
(c) Coupling: between 2 recessive alleles, repulsion: between dominant and recessive alleles
(d) Both (a) and (c)

11. Beard development in man is

(a) Sex linked character
(b) Sex limited character
(c) Both (a) and (b)
(d) None of the above

12. Base analogues used as mutagens

(a) 5 bromouracil
(b) 3 bromouracil
(c) Bromouracil
(d) 2 bromouracil

13. Following is correct related to the expression of sex-linked characters

(a) Recessive gene can express in homogametic conditions
(b) Recessive allele can express in heterogametic conditions
(c) Recessive allele can express in homogametic conditions
(d) Both (a) and (b)

14. Genes of sex-linked characters are present on

(a) X chromosome
(b) Y chromosome
(c) Autosomes
(d) All of the above

15. Criss cross inheritance is

(a) Inheritance of sex-linked gene from grandmother to granddaughter through daughter

(b) Inheritance from grandfather to grandson through daughter

(c) Inheritance from grandfather to grandson through son

(d) Both (b) and (c)

16. Cytoplasmic inheritance was first reported in ______ for the trait______

(a) Mirabilis, for leaf colour

(b) Mirabilis, for flower colour

(c) Mirabilis, for seed colour

(d) All of the above

17. Gene for gene hypothesis was reported by

(a) Rimpu associated with linseed rust

(b) Jones associated with linseed smut

(c) Flor associated with linseed rust

(d) Donald associated with linseed smut

18. Tendency of one crossover to reduce the chance of another crossover is

(a) Linkage

(b) Coincidence

(c) Interference

(d) None of the above

19. Colour blind carrier girl x colour bind boy so progeny that are colour blind

(a) 100 per cent

(b) 75 per cent

(c) 25 per cent

(d) 50 per cent

20. Pureline is

(a) Homogenous population of cross-pollinated plant

(b) Homogenous population of often cross-pollinated plant

(c) Homogenous population of homozygous self-pollinated plant

(d) Homogenous population of heterozygous self-pollinated plant

21. Method of improving both self- and cross-pollinated species is

(a) Pedigree method

(b) Clonal selection

(c) Mass selection

(d) Bulk breeding methods

22. Following is correct

(a) Back cross maybe a test cross

(b) Test cross maybe a back cross

(c) Both (a) and (b)

(d) None of the above

23. Oldest method of crop improvement is

(a) Mass method
(b) Mass selection
(c) Pureline selection
(d) Recurrent selection

24. Progeny of a single plant obtained through asexual reproduction is

(a) Mulitline
(b) Clone
(c) Pedigree
(d) None of the above

25. Ideotype concept was given by

(a) Donald
(b) Johanson
(c) Jonnes
(d) Flor

26. Main centers and subcenters of crop diversity identified are

(a) 3, 8
(b) 8, 3
(c) 4, 5
(d) 6, 8

27. First interspecific hybrid developed by

(a) Thomas fairchild
(b) Rimpu
(c) Jones
(d) Flor

28. Concept of resistance was given by

(a) Vander plank
(b) Waldeyer
(c) Strasburger
(d) None of the above

29. Double cross in maize was done for the first time by

(a) Johanson
(b) Jones
(c) Robert
(d) None of the above

30. U 's triangle was given by

(a) Nagaharu
(b) Mendel
(c) Jones
(d) Fairchild

31. Mass pedigree method was proposed by

(a) Shull
(b) Harris
(c) Harrington
(d) Muller

32. Mutation was first defined by

(a) Muller
(b) Hugo de vries
(c) Both (a) and (b)
(d) Bateson

33. Linkage theory was given by

(a) Muller
(b) Morgan
(c) Vavilon
(d) Bateson

34. Cytoplasmic inheritance is governed by

(a) Plasma genes (b) Mitochondrial genes

(c) Nuclear genes (d) All of the above

35. DT 46 is the variety of

(a) Wheat (b) Rye

(c) Triticale (d) None of the above

36. Disease resistance are mainly developed by

(a) Mass selection (b) Mass breeding

(c) Recurrent selection (d) Back cross method

37. Sex linked character is

(a) White eye in drosophila (b) Colour blindness

(c) Hemophilia (d) All of the above

38. Cytoplasmic inheritance was reported by

(a) Waldeyer (b) Carson

(c) Correns (d) Muller

39. Method used for development of inbred lines is

(a) Pedigree method (b) Bulk breeding method

(c) Back cross method (d) None of the above

40. Crop ideotype was first reported while working with

(a) Rice (b) Wheat

(c) Maize (d) Rye

41. Per capita Forest land availability of the country is

(a) 0.02 hectare (b) 0.05 hectare

(c) 0.08 hectare (d) 0.1 hectare

42. Khadi Prakritik Paint is made up of

(a) green plants (b) cow urine

(c) cow dung (d) All of the above

43. Total micro irrigation area of India is

(a) 10.8 Mha (b) 12.9 Mha

(c) 15.6 Mha (d) 8.4 Mha

44. MITRA stands for

(a) Micro Investment Textile Parks

(b) Mega Investment Textile Parks

(c) Mega Investment Trading Parks

(d) None of the above

45. Contribution of food processing industry to agricultural sector is

(a) 15 per cent (b) 12 per cent

(c) 9 per cent (d) 11 per cent

46. Operation launched to rescue Indian citizens from Ukraine

(a) Operation Yamuna (b) Operation Ganga

(c) Operation Raahat (d) Operation Bharat

47. First Indian to be named after Boltzmann Medal recently

(a) Ashok Sen (b) Piara Singh Gill

(c) Deepak Dhar (d) Meghnad Saha

48. National Protein Day is celebrated on

(a) 12th February (b) 29th February

(c) 27th February (d) 10th March

49. Which country along with India conducted Ex Dharma Guardian Military exercise

(a) Sri Lanka (b) Japan

(c) France (d) USA

50. National Security Day is celebrated on

(a) 4th March (b) 1st March

(c) 10th February (d) 15th March

ANSWER KEY

1	2	3	4	5	6	7	8	9	10	11	12	13	14	15
(b)	(b)	(b)	(c)	(c)	(d)	(d)	(d)	(c)	(d)	(b)	(a)	(d)	(a)	(b)
16	17	18	19	20	21	22	23	24	25	26	27	28	29	30
(a)	(c)	(c)	(d)	(c)	(c)	(a)	(b)	(b)	(a)	(b)	(a)	(a)	(b)	(a)
31	32	33	34	35	36	37	38	39	40	41	42	43	44	45
(c)	(b)	(b)	(a)	(c)	(d)	(d)	(c)	(a)	(b)	(c)	(c)	(b)	(b)	(c)
46	47	48	49	50										
(b)	(c)	(c)	(b)	(a)										

Chapter 15

Seed Technology

1. **The superiority of F1 in one or more character over its parents is known as**
   (a) Heterosis (b) Hybrid vigour
   (c) Hybrid vitality (d) Both (a) and (b)

2. **Complete self-pollination is favoured by**
   (a) Geitenogamy (b) Chasmogamy
   (c) Cleistogamy (d) All of the above

3. **Variation in genetic material occurs in**
   (a) Geitenogamy (b) Chasmogamy
   (c) Cleistogamy (d) All of the above

4. **Phenomenon where one gene affects the expression of other gene**
   (a) Euphenics (b) Eugenics
   (c) Epistasis (d) None of the above

5. **Total variability present within and among all living organisms is**
   (a) Variation (b) Biodiversity
   (c) Bioversatality (d) None of the above

6. **Mixture of genotypes from several sources maintained in bulk from one generation to the next**
   (a) Synthetic variety (b) Composite variety
   (c) Multiline (d) Pureline

7. The following causes inbreeding depression

(a) Repeated selfing (b) Repeated sub mating
(c) Repeated cross pollination (d) Both (a) and (b)

8. In pseudo heterosis, superiority of F1 is in

(a) Adaptation (b) Yield
(c) Vegetative growth (d) Both (a) and (b)

9. Top cross involves cross between

(a) Inbred line x open pollinated variety
(b) Outbread line x often cross-pollinated variety
(c) Pureline x cross pollinated variety
(d) All of the above

10. Geitenogamy is seen in

(a) Rice (b) Date palm
(c) Papaya (d) Maize

11. Maintainer line is

(a) A line, fertile (b) B line, fertile
(c) A line, sterile (d) B line, sterile

12. R line x A line yields

(a) Male fertile hybrid (b) Male sterile hybrid
(c) Both sterile and fertile in equal ratios (d) None of the above

13. Genetic purity of nucleus seed is

(a) 98 per cent (b) 97 per cent
(c) 99 per cent (d) 100 per cent

14. Progeny of nucleus seed is

(a) Nucleus seed itself (b) Breeder seed
(c) Foundation seed (d) Registered seed

15. Progeny of breeder seed is. With the colour tag.

(a) Nucleus seed, no colour
(b) Foundation seed, golden yellow
(c) Foundation seed, white colour
(d) Registered seed, blue colour

16. The following seed is generally omitted in India

(a) Breeder seed
(b) Foundation seed
(c) Registered seed
(d) Certified seed

17. Approved by state seed agency

(a) Registered seed
(b) Foundation seed
(c) Certified seed
(d) Nucleus seed

18. Isolation distance is maximum for

(a) Maize
(b) Wheat
(c) Cotton hybrid
(d) Pearl millet hybrid

19. Quickest method of viability test is

(a) Potassium permanganate method
(b) Tetrazolium chloride test
(c) Grodex test
(d) All of the above

20. Grodex is a brand name of

(a) Triphenyl trizolium bromide
(b) Triphenyl trizolium chloride
(c) Triphenyl tetrazolium chloride
(d) Triphenyl tetrazolium bromide

21. Qualitative test is

(a) Potassium permanganate method
(b) Tetrazolium chloride test
(c) Grodex test
(d) All of the above

22. Biological test is

(a) Potassium permanganate method
(b) Tetrazolium chloride test
(c) Grodex test
(d) All of the above

23. Real value of seed is

(a) Purity per cent x germination per cent
(b) Purity per cent/germination per cent
(c) Purity per cent + germination per cent
(d) None of the above

24. Mother seed is

(a) Breeder seed
(b) Nucleus seed
(c) Foundation seed
(d) Registered seed

25. Seed meant for commercial cultivation by farmers is

(a) Breeder seed
(b) Foundation seed
(c) Registered seed
(d) Certified seed

26. National seed corporation was setup in

(a) 1963
(b) 1968
(c) 1966
(d) 1993

27. National seed act was passed in the year

(a) 1963
(b) 1966
(c) 1968
(d) 1948

28. Seed available in limited quantity is

(a) Certified seed
(b) Nucleus seed
(c) Foundation seed
(d) All of the above

29. Seeds of cereals are

(a) Orthodox seeds which remain viable at low temperature and moisture
(b) Orthodox seeds which do not remain viable at low temp. And moisture
(c) Recalcitrant seeds which remain viable at lo temp. And moisture
(d) Recalcitrant seeds which do not remain viable at low temp. And moisture

30. Central laboratory of Agmark is at

(a) Mumbai
(b) Nagpur
(c) Delhi
(d) Chennai

31. Following is correct combination

(a) Seed index: weight of 1000 seeds, test weight: weight of 100 seeds
(b) Seed index: weight of 100 seed, test weight: weight of 1000 seeds
(c) Both seed index and test weight are synonymous
(d) None of the above

32. Highest and lowest germination percentages of fields crops are respectively

(a) Maize (90), Cotton (40)
(b) Rice (90), Pearl millet (40)
(c) Maize (90), Cotton (65)
(d) Wheat (90), Cotton (50)

33. Seed plot technique of potato was given by

(a) Pushkarnath
(b) Ramakrishna
(c) Khush
(d) None of the above

34. Seed was declared as

(a) Luxurious commodity (b) Non-essential commodity

(c) Essential commodity (d) None of the above

35. Impurity by seeds of other crop species permitted in foundation seed is

(a) 0.1 per cent (b) 1 per cent

(c) 10 per cent (d) 2 per cent

36. Seed quality index measures

(a) Vigour (b) Viability

(c) Both (a) and (b) (d) Strength of seed coat

37. Strongest dormancy breaker is

(a) PNO_3 (b) KNO_3

(c) PPO_3 (d) KPO_3

38. Seed plot technique is for developing

(a) High yielding potatoes (b) Stress tolerant potatoes

(c) Disease free potatoes (d) All of the above

39. First synthetic variety was developed in

(a) Wheat (b) Maize

(c) Rice (d) Cotton

40. Dormancy breaker in potato is

(a) Thiourea (b) Urea

(c) Thionitrate (d) All of the above

41. State in India using Automated Weather Stations

(a) Maharashtra (b) Karnataka

(c) Telangana (d) Kerala

42. First Pashu Kisan Credit cards were given in

(a) Haryana (b) Hyderabad

(c) Chennai (d) Bhubaneshwar

43. National Sugar Institute of India will establish sugar institute at

(a) USA (b) Nigeria

(c) Canada (d) China

44. India's largest semen station is at

(a) Bihar
(b) Uttar Pradesh
(c) Andhra Pradesh
(d) Gujarat

45. First gender park was setup at

(a) Kerala
(b) Tamil Nadu
(c) Karnataka
(d) Telangana

46. Which Union Ministry implements the Mahatma Gandhi National Rural Employment Guarantee Scheme (MGNREGS)?

(a) Ministry of Labour and Employment
(b) Ministry of Rural Development
(c) Ministry of Social Justice and Empowerment
(d) Ministry of Law and Justice

47. Which country has been hit by extra-ordinary rains called 'Rain Bombs'?

(a) USA
(b) Australia
(c) Indonesia
(d) Sri Lanka

48. The 'Yilan' Crater, the largest impact crater on Earth, has been discovered in which country?

(a) Russia
(b) China
(c) Australia
(d) Japan

49. What does 'U' stand for in the Logistics Platform ULIP, announced in the Union Budget 2022-23?

(a) Unit
(b) Unified
(c) Universal
(d) Unique

50. DIKSHA, SWAYAM and NISHTHA are initiatives of which Union Ministry?

(a) Ministry of MSME
(b) Ministry of Education
(c) Ministry of Skill Development
(d) Ministry of Environment, Forest and Climate Change

ANSWER KEY

1	2	3	4	5	6	7	8	9	10	11	12	13	14	15
(d)	(c)	(b)	(c)	(b)	(a)	(d)	(c)	(a)	(d)	(b)	(a)	(d)	(b)	(c)
16	17	18	19	20	21	22	23	24	25	26	27	28	29	30
(c)	(c)	(d)	(b)	(d)	(a)	(b)	(a)	(c)	(d)	(a)	(b)	(b)	(a)	(b)
31	32	33	34	35	36	37	38	39	40	41	42	43	44	45
(b)	(c)	(a)	(c)	(a)	(a)	(b)	(c)	(b)	(a)	(a)	(a)	(b)	(a)	(a)
46	47	48	49	50										
(b)	(b)	(b)	(b)	(b)										

PLANT PATHOLOGY

Chapter 16

Introduction of Plant Pathology, Historical Background, History of Plant Pathology in India, Principles of Plant Pathology, Types of Diseases

1. **Founder of mycology?**
   (a) Linnaeus (b) Anton de bary
   (c) P A Macheli (d) E J Butler
2. **First identified bacterial disease of plant "fire blight of apple" was discovered by**
   (a) Adolf Mayer (b) E F Smith
   (c) M K Patel (d) T J Burril
3. **In India green year disease of sorghum was first identified by**
   (a) Butler (b) Mundkar
   (c) Pathak (d) Kirtikar
4. **Phytoplasma can cause plant disease was first reported by?**
   (a) Diener (b) Doi *et al.*
   (c) Kassanis (d) Herelle
5. **Satellite virus was first discovered by**
   (a) Kassanis (b) Diener
   (c) Bawden (d) Johnson

6. Ireland famine is due to

(a) Early blight of potato (b) Wart disease of potato

(c) Late blight of potato (d) PVX

7. Who proved first time that fungi could cause plant disease?

(a) Micheli (b) Tillet

(c) Prevost (d) Millardet

8. Sexuality of rust fungi was first discovered by?

(a) Anton de bary (b) Stakman

(c) Cargie (d) Butler

9. Transduction of bacteria was first demonstrated by?

(a) Zinder and Lederberg (b) Grifith

(c) Herelle (d) Muller

10. The first popular classical fungicide is?

(a) Burgundy mixture (b) Bordeaux mixture

(c) Cheshunt compound (d) Dithane M 45

11. The first antibiotic was?

(a) Penicillin (b) Streptomycin

(c) Tetracycline (d) Erythromycin

12. Bordeaux mixture was discovered by?

(a) Prevost (b) Tillet

(c) Anton de Bary (d) Millardet

13. The first fungus known to man is?

(a) Amanita (b) Puccinia

(c) Phytopthora (d) Albugo

14. Who introduced the term Viroid?

(a) Doi *et al.* (b) Kassanis

(c) Diener (d) Symons

15. The great Irish famine was during

(a) 1940-41 (b) 1945-1946

(c) 1845-1846 (d) 1841-1842

16. Father of Virology?

(a) Beijernick (b) Ivanowski

(c) Bawden (d) Mayer

17. Father of bacteriology?

(a) Robert Koch (b) Leeuwenhock

(c) Burrill (d) Smith

18. Father of phytobacteriology?

(a) Koch (b) Leewenhock

(c) Burrill (d) Smith

19. Father of morden Indian phytopathology?

(a) K C Mehta (b) EJ Butler

(c) Dastur (d) Mundkar

20. Father of wheat rust disease in India?

(a) K C Mehta (b) EJ Butler

(c) Dastur (d) Mundkar

21. Who has established Indian phytopathological society at new delhi?

(a) KC Mehta (b) EJ Butler

(c) Dastur (d) Mundkar

22. Gene for gene hypothesis of disease resistance is proposed by?

(a) Burrill (b) Flor

(c) Kassanis (d) Shepherd

23. Father of pure culture technique?

(a) Brefeld (b) Bary

(c) Prevost (d) Linnaeus

24. Who gave the concept of Vertical and horizontal resistance in plant?

(a) Shepherd (b) Diener

(c) Koch (d) Vanderplank

25. Coffee rust occurred in Sri Lanka during

(a) 1845 (b) 1843

(c) 1867 (d) 1945

26. When a disease is constantly present from year to year in a particular part of country or state?

(a) Endemic (b) Epidemic

(c) Sporadic (d) Pandemic

27. A disease which occurs widely in severe but periodically

(a) Endemic (b) Epidemic

(c) Sporadic (d) Pandemic

28. Disease which occurs at very irregular interval and location

(a) Endemic (b) Epidemic

(c) Sporadic (d) Pandemic

29. When a disease prevalent throughout the country or world is known as?

(a) Endemic (b) Epidemic

(c) Sporadic (d) Pandemic

30. Example of Endemic disease is?

(a) Red rot of sugarcane (b) Green ear of bajra

(c) Late blight of potato (d) Wart of potato

31. Examples of Autoecious rust is?

(a) Linseed rust (b) Sunflower rust

(c) Safflower rust (d) All of the above

32. Example of heteroecious rust

(a) Linseed rust (b) Safflower rust

(c) Sunflower rust (d) Wheat rust

33. Sooty or charcoal like powder is characteristics symptom of?

(a) Rust (b) Smut

(c) Downy mildew (d) Powdery mildew

34. Drying of twigs and branches of plant from the top to backwards is known as?

(a) Die back (b) Downy mildew

(c) Powdery mildew (d) None of the above

35. Death of tissues or cells is known as?

(a) Necrosis (b) Blight

(c) Chlorosis (d) Wilt

36. Increase the number of cells is known as?

(a) Hyperplasia (b) Hypertrophy

(c) Canker (d) Scab

37. Increase in the size of the cell is known as?

(a) Hyperplasia (b) Hypertrophy

(c) Canker (d) Scab

38. which one disease is beneficial for the growers?

(a) Late blight of potato (b) Fire blight of apple

(c) Loose smut of wheat (d) Colour breaking of tulip

39. Which one disease of wheat causes restriction in export of wheat from India?

(a) Kernal bunt (b) Karnal bunt

(c) Loose smut (d) Flag smut

40. Rishitin is a phytoalexin and produced by?

(a) Tomato (b) Tobbacco

(c) Rice (d) Potato

41. First pollution polluted disease in plant?

(a) Black tip of mango (b) Silvery leaf disease

(c) Khaira disease of rice (d) white tip of paddy

42. Domestic quarantine exists on?

(a) Rice blast (b) Red rot of sugarcane

(c) Late blight of potato (d) Wart of potato

43. The organisms lacking a well-defined cell wall is termed as?

(a) Fungi (b) Protozoa

(c) Eukaryotes (d) Prokaryotes

44. Exclusion of plant disease by legislation is known as?

(a) Plant quarantine (b) Disease resistance

(c) Cultural control (d) Biological control

45. Conidia and sporangia are spores produced by?

(a) Viruses (b) Fungi

(c) Bacteria (d) Phytoplasma

45. Which of the following is not correctly matched

(a) Mustard- white rust (b) Paddy-Brown rust

(c) Wheat-Red rust (d) Groundnut – Tikka

46. Tungro disease of rice is spread by

(a) *Nephotettix virescense* (b) *Sogatella fucifera*

(c) *Nilaparvatha leugens* (d) *Thrips tabaci*

47. The term hypersensitive response is given by

(a) Stakmann (b) Erickson

(c) H.M. Ward (d) W.A. Orton

48. Soil microorganism are most active at

(a) 18-20° (b) 26-28°

(c) 30-30° (d) 34-36°

49. VAM is

(a) Bacteria (b) Fungi

(c) Virus (d) Algae

50. Which of the following is not a disease of apple

(a) Bird eye (b) Powdery mildew

(c) Downy mildew (d) Bunchy top

ANSWER KEY

1	2	3	4	5	6	7	8	9	10	11	12	13	14	15
(c)	(d)	(a)	(b)	(a)	(c)	(c)	(b)	(a)	(b)	(a)	(d)	(a)	(c)	(b)
16	17	18	19	20	21	22	23	24	25	26	27	28	29	30
(a)	(b)	(d)	(b)	(a)	(d)	(b)	(a)	(d)	(c)	(a)	(b)	(c)	(d)	(d)
31	32	33	34	35	36	37	38	39	40	41	42	43	44	45
(d)	(d)	(b)	(a)	(a)	(a)	(b)	(d)	(b)	(d)	(b)	(d)	(d)	(a)	(b)
46	47	48	49	50										
(a)	(d)	(d)	(b)	(d)										

Chapter 17

Mycology, Fungicides, Symptoms of Diseases Caused due to Fungi, Fungi Structure, Life Cycle, Asexual and Sexual Fruiting Bodies, Orders, Family Characters

1. **Coenocytic mycelium found in?**
   (a) Oomycetes (b) Ascomycetes
   (c) Baidiomycetes (d) Deuteromycetes
2. **Gama particles are found in fungal division of**
   (a) Ascomycotina (b) Basidiomycotinna
   (c) Deuteromycotina (d) Mastigomycotina
3. **Example of Polymorphic fungus is?**
   (a) *Puccinia graminis tritici* (b) *Albugo candida*
   (c) *Alternaria solani* (d) *Erysiphe polygony*
4. **The term mycorrhizae means**
   (a) Mushroom like (b) Fungus roots
   (c) Rhizobium like (d) Plant roots
5. **Example of well proven Antifungal antibiotic is?**
   (a) Penicillin (b) Streptomycin
   (c) Amphoterein (d) Tetramycin

6. **Bioluminescence property is found in**
   (a) *Amanita muscaria* (b) *Armillaria mellea*
   (c) *Rhizopus stolonifer* (d) *Lentinus edodes*

7. **A naked multinucleate motile mass of protoplasm of fungi is known as?**
   (a) Flagella (b) Mycelium
   (c) Plamodium (d) Pili

8. **The well-known hallucinogenic mushroom is**
   (a) *Psilocybe mexicana* (b) *Pleurotus sojor*
   (c) *Agaricus bisporus* (d) *Volvariella volvacea*

9. **Charcoal rot is caused by**
   (a) *Botrytis* spp. (b) *Macrophomina* spp.
   (c) *Taphrina* spp. (d) *Sclerotinia* spp.

10. **Apple Scab is caused by**
   (a) *Erwinia amylovora* (b) *Synchytrium endobioticum*
   (c) *Spilocaea pomi* (d) *Synchytrium taraxaci*

11. **Powdery mildew fungi are identified by presence of**
   (a) Conidia (b) Sporangiospores
   (c) Zoospores (d) Appendages

12. **Cleistothecium is a**
   (a) Open ascocarp (b) Closed ascocarp
   (c) Cup shaped (d) Flask shaped

13. **Red onion is resistance to smudge due to presence of?**
   (a) Silicon (b) Protocol
   (c) Methyl amine (d) Protocatechuic acid

14. **Mycoparasitism is the phenomenon between**
   (a) Fungi and bacteria (b) Fungi and Fungi
   (c) Bacteria and virus (d) Fungi and virus

15. **Most endomycorrhyzae are known as**
   (a) VA Mycorrhyzae (b) Orchidmycorrhizae
   (c) Criphytemycorrhizae (d) Terrestrialmycorrhizae

16. **The Powdery mildews are under order**
   (a) Taphrinales (b) Erysiphales
   (c) Endomycetales (d) Pezizals

17. Hartig nets are found in

(a) Ectomycorhizae
(b) Endomycorhizae
(c) Oomycetous Fungi
(d) Deuteromycotina fungi

18. The fungus having largest host range?

(a) Phytophthera
(b) Rhizoctonia
(c) Fusarium
(d) Sclerotium

19. The term Phytophthora means

(a) Plant destroyer
(b) Root rot
(c) Phyton like
(d) Potato rot

20. Foul smell of wheat grain infected by Tilletia caries and *T. foetida* is due to

(a) LSD
(b) Aflatoxin
(c) Mycotina
(d) Chlymydospores

21. *Taphrina deformans* causes

(a) Tomato leaf curl
(b) Potato leaf curl
(c) Chillies leaf curl
(d) Peach leaf curl

22. *Taphrina maculans* causes disease

(a) Brown spot of turmeric
(b) Tomato leaf curl
(c) Peach leaf curl
(d) Chilli leaf curl

23. Albugo is

(a) Predator
(b) Saprophyte
(c) Facultative parasite
(d) Obligate parasite

24. Linseed rust is caused by?

(a) *Uromyces fabae*
(b) *Physagmidium pisi*
(c) *Puccinia graminis*
(d) *Melampsora lini*

25. In plains of India which rust appear first

(a) Black rust
(b) Yellow rust
(c) Orange rust
(d) None of the above

26. Largest subdivision of fungi

(a) Deuteromycotina
(b) Basidiomycotina
(c) Mastigomycotina
(d) Ascomycotina

27. Two chains of fungal spores remain contact with a ped known as

(a) Isthmus (b) Conidiophore

(c) Hymenium (d) Croizer

28. Smut of wheat is effectively controlled by

(a) Vitavax (b) Bavistin

(c) Thirum (d) Benlate

29. Which of the following Fungi is a poisonous mushroom?

(a) Agaricus (b) Volvariella

(c) Pleurotus (d) Amanita

30. Foot rot of papaya is caused by

(a) *Perenospora parasitica* (b) *Pythium aphanidermatum*

(c) *Phytophthera parasitica* (d) *Phytophthera palmivora*

31. Which fungal genus is oppertunistic pathogen of human?

(a) Phytophthera (b) Puccinia

(c) Rhizopus (d) Alternaria

32. Kingdom Chromista is composed of

(a) Chitin (b) Glucan

(c) Levans (d) Cellulose

33. Uncinula causes plant disease

(a) Downy mildew (b) Powdery mildew

(c) Blight (d) Rust

34. Which fungus suppresses various pathogenic fungi present in soil?

(a) Pythium (b) Fusarium

(c) Macrophomina (d) Trichoderma

35. Powdery mildews are controlled by spraying with

(a) Copper fungicide (b) Mercury fungicide

(c) Sulphur (d) None of the above

36. Red rot of sugarcane is caused by?

(a) Fusarium (b) Alternaria

(c) Colletotrichum (d) Pyricularia

37. Loose smut of wheat is a disease known as

(a) Internally seed borne (b) Externally seed borne

(c) Air borne (d) Soil borne

38. Fungi which can grow only on living host plants are known as

(a) Facultative parasites (b) Obligate parasites

(c) Saprophyte (d) None of the above

39. White rust of crucifer is caused by?

(a) Albugo (b) Puccinia

(c) Urocystis (d) Peronospora

40. Late blight of potato is caused by?

(a) *Alternaria solani* (b) *Phytophthera palmivora*

(c) *Phytophthera parasi* (d) *Phytophthera infestance*

41. Which fungus is known as "honey or boot lace" Fungus?

(a) Armillaria mella (b) Amanita muscaria

(c) Coprinus comatus (d) *Psilocybe*

42. Nutrition of Erysiphe is

(a) Ectoparasite (b) Endoparasite

(c) Both (d) None of the above

43. Which of the fungus is known as "fly fungus"

(a) *Amanita verna* (b) *A. Phalloides*

(c) *A. partherina* (d) *A. muscaria*

44. Auto digestion is commonly found in which fungus

(a) Amanita (b) Psilocybe

(c) Coprinus (d) Lactarius

45. Largest conidia size are found in ?

(a) Rusts (b) Smuts

(c) Powdery mildews (d) Downy mildew

46. Sexual spore produced by fungi causing downy mildew diseases

(a) Zygospore (b) Teliospore

(c) Oospore (d) Zoospore

47. Wilt disease of banana caused by *Ralstonia solanacearum* is called as

(a) Panama wilt of banana
(b) Slow decline of banana
(c) Moko disease of banana
(d) Sigatoka disease

48. Bursting of Pseudostem in banana is observed in plants suffering from

(a) Bunchy top
(b) Panama wilt
(c) Root-knot nematode disease
(d) Banana streak virus

49. Downy mildew is caused by the member of the family

(a) Pythiaceae
(b) Albuginaceae
(c) Perenpsporaceae
(d) Albuginaceae

50. Nylon netting of Tomato nursery is practiced for the management of

(a) Bacterial wilt
(b) Early blight
(c) Leaf curl
(d) Late blight

ANSWER KEY

1	2	3	4	5	6	7	8	9	10	11	12	13	14	15
(a)	(d)	(a)	(b)	(c)	(b)	(c)	(a)	(b)	(c)	(d)	(b)	(d)	(b)	(a)
16	17	18	19	20	21	22	23	24	25	26	27	28	29	30
(b)	(a)	(d)	(a)	(d)	(d)	(a)	(d)	(d)	(b)	(d)	(a)	(a)	(d)	(b)
31	32	33	34	35	36	37	38	39	40	41	42	43	44	45
(c)	(b)	(b)	(d)	(c)	(c)	(a)	(b)	(a)	(d)	(a)	(a)	(d)	(c)	(c)
46	47	48	49	50										
(c)	(c)	(b)	(b)	(b)										

Chapter 18

Bacteriology, Virology, Molecular Interaction between Host and Pathogen, Molecular Techniques, Principles of Plant Disease Management

1. **Bacteria that infect bacteria are known as?**
   (a) Bacteriophage (b) Cyanophage
   (c) Bdellovibrio (d) L- form bacteria

2. **In gram staining, Clavibacter shows**
   (a) Red colour (b) Violet colour
   (c) Blue colour (d) Green colour

3. **L- form bacteria are**
   (a) Pathogenic (b) Saprophytic
   (c) Anaerobic (d) Cell wall lacking

4. **The test which confirm bacterial infection in plant is?**
   (a) Iodine test (b) Ooze test
   (c) Cobalt test (d) All of the above

5. **The number of ring present in gram positive bacterial flagella are**
   (a) 4 (b) 2
   (c) 3 (d) 1

6. **The number of rings present in gram negative bacterial flagella is**
   (a) 4 (b) 3
   (c) 2 (d) 1

7. **Most plant pathogenic bacteria are**
   (a) Gram negative (b) Gram positive
   (c) Non flagellable (d) Cell wall lacking

8. **Bacterial flagellum is consist of**
   (a) Meurin (b) Glycoprotein
   (c) Nucleic acid (d) Flagellin

9. **Which one of the following is a fastidious vascular bacteria pathogen**
   (a) Psudomonas (b) Erwinia
   (c) Xanthomonas (d) Xylalla

10. **Bacteria chromosome is composed of**
   (a) ss DNA (b) ds DNA
   (c) ss RNA (d) ds RNA

11. **Reseve food material of bacteria is known as**
   (a) Glycogen (b) glucan
   (c) Hydroxybutyrate (d) Cellulose

12. **Flagellar arrangements of erwinia is known as**
   (a) Monotrichus (b) Amphitrichous
   (c) Lophotrichous (d) Peritrichus

13. **Bacteria endospores are stained with**
   (a) Fuchsin (b) Nigrocin
   (c) Malachite green (d) Gram stain

14. **Division of gram-positive bacteria is**
   (a) Tenericutes (b) Gracilicutes
   (c) Firmicutes (d) Molicutes

15. **The division of gram-negative bacteria**
   (a) Tenericutes (b) Gracilicutes
   (c) Firmicutes (d) Molicutes

16. **Bacteria are multiplying by**
   (a) Budding (b) Parasexuality
   (c) Binary fission (d) Fragmentation

17. Symptoms are produced when the population of plant pathogenic bacteria reaches about

(a) 10 power 4 CFU/ml
(b) 10 power 3 CFU/ml
(c) 10 power 5 CFU/ml
(d) 10 power 7 CFU/ml

18. Bdellovibrio kills

(a) Gram negative bacteria
(b) Gram positive bacteria
(c) Phytoplasma
(d) Viruses

19. Bacteria are

(a) Eukaryotic
(b) Prokaryotic
(c) Animals
(d) Plants

20. Most plant pathogenic bacteria are

(a) Coccus
(b) Rod shaped
(c) Filamentous
(d) Pleomorphic

21. Corn stunt disease is caused by?

(a) Spiroplasma
(b) Virus
(c) Viroid
(d) None of the above

22. Spiroplasma is present with in

(a) Xylem
(b) Phloem sieve tubes
(c) Root cortex
(d) Apical meristem tissue

23. Mycoplasma are

(a) Rod shaped
(b) Spherical
(c) Gram positive
(d) Gram negative

24. The bacterial genus caused maximum plant disease is

(a) Xanthomonas
(b) Pseudomonas
(c) Erwinia
(d) Agrobacterium

25. Bacteria are

(a) Haploid
(b) Diploid
(c) Tetraploid
(d) None of the above

26. Bacterial sex factor is also known as

(a) Col factor
(b) Fertility or F factor
(c) Cryptic plasmid
(d) R factor

27. The extra chromosomal elements of bacteria are known as

(a) Genophores (b) Chromatids

(c) Plasmids (d) F- factor

28. On culture media mycoplasma colonies appear as

(a) Fried egg appearance (b) Rod shaped

(c) Helical (d) Spiral

29. Extra chromosomal elements of bacteria are known as

(a) Genophores (b) Chromatids

(c) Plasmids (d) F factor

30. Mycovirus contains

(a) ss RNA (b) ss DNA

(c) ds DNA (d) ds RNA

31. Viruses that infect blue green algae are called as

(a) Bacteriophage (b) Cyanophage

(c) Mycovirus (d) Viroid

32. Viruses that kill some gram negative bacteria are known as

(a) Virusoids (b) Virions

(c) Bdellovibrio (d) Bacteriophages

33. The first plant viroid disease reported was

(a) Coconut cadang cadang (b) Citrus exocortis

(c) Potato spindle tuber (d) Hop stunt

34. Toxoptera citricida is vector of

(a) Citrus tristeza (b) Citrus exocortis

(c) Citrus greening (d) Citrus scab

35. The first identified plant viral disease was

(a) Citrus tristeza (b) Tobacco mosaic

(c) Cauliflower mosaic (d) Tomato leaf curl

36. Oldest known viral disease of plant is

(a) Tulip colour breaking (b) Tobacco mosaic

(c) Potato mosaic (d) Cauliflower mosaic

37. Viroid are

(a) ss DNA (b) ds RNA

(c) ds DNA (d) Circular naked RNA

38. Virus particle with both nucleic acid and protein coat is known as

(a) Virion (b) Viroid

(c) Capsid (d) All of the above

39. Tomato leaf curl virus is transmitted by

(a) Myzus persici (b) Bemicia tabaci

(c) Thrips tabaci (d) Xiphinema

40. Cell to cell movement of virus takes place through

(a) Xylem (b) Plastids

(c) Plasmodesmata (d) Chlorophyll

41. Cacao swollen shoot virus is transmitted by

(a) Mealybug (b) White fly

(c) Aphids (d) Thrips

42. The bimodal transmission is found in case of

(a) TMV (b) CaMV

(c) CMV (d) PVY

43. Rice tungro spherical virus have genome of

(a) ss RNA (b) ss DNA

(c) ds RNA (d) ds DNA

44. Rice tungro baciliform virus has the genome of

(a) ss RNA (b) ss DNA

(c) ds RNA (d) ds DNA

45. The term virus means

(a) Very small (b) Poison

(c) Toxin (d) Animacules

46. In the peptidoglycon of the bacterial cell wall the two sugar molecules are connected by

(a) Peptide Cross links (b) Glycosidic bond

(c) H-bond (d) S-H bond

47. Porins are present in bacterial cell wall component

(a) LPS layer
(b) EPS layer
(c) Cytoplasmic membrane
(d) Plasma membrane

48. Bacterial cell wall is made up of

(a) Peptidoglycon
(b) Chitin
(c) Flagellin
(d) Lipopolysaccharide

49. The genetic material in a bacterial cell is

(a) A nucleoid region
(b) Definite nucleus
(c) Mitochondria
(d) Mesosome

50. Internally seed borne bacterial disease

(a) Black rot of cabbage
(b) Soft rot of carrot
(c) Pierce disease of Grapes
(d) Citrus canker

ANSWER KEY

1	2	3	4	5	6	7	8	9	10	11	12	13	14	15
(c)	(b)	(d)	(b)	(b)	(a)	(a)	(d)	(d)	(b)	(c)	(d)	(c)	(c)	(b)
16	17	18	19	20	21	22	23	24	25	26	27	28	29	30
(c)	(d)	(a)	(b)	(b)	(a)	(b)	(d)	(b)	(a)	(b)	(c)	(a)	(c)	(d)
31	32	33	34	35	36	37	38	39	40	41	42	43	44	45
(b)	(d)	(c)	(a)	(b)	(a)	(d)	(a)	(b)	(c)	(a)	(b)	(a)	(b)	(b)
46	47	48	49	50										
(b)	(a)	(a)	(a)	(a)										

Chapter 19

Epidemiology, Crop Diseases of Rice, Wheat, Barley, Pearl Millet, Maize, Gram, Sorghum, Arhar, Soyabean, Groundnut, Mustard, Sugarcane, Potato, Sesame, Cotton, Tomato, Brinjal, Capsicum, Crucifers, Banana, Citrus, Cumin, Cardamom

1. **Richman's disease and poorman's apple respectively are**
   (a) Rice blb, apple
   (b) Papaya ring spot virus, jackfruit
   (c) Rice blast, guava
   (d) Rice blast, jackfruit

2. **Most severely effecting pathogen in jeera is**
   (a) *Alternaria burnsii*
   (b) *Colletotrichum cumini*
   (c) *Fusarium vasinfectum cumini*
   (d) *Fusarium oxysporum cumini*

3. **Marble disease is associated with**
   (a) Katte disease of clove
   (b) Cardamom mosaic virus
   (c) Both (a) and (b)
   (d) Only (b)

4. **Premature opening of bolls is associated with**
   (a) Angular leaf spot caused by alternaria
   (b) Black arm caused by xanthomonas
   (c) Both (a) and (b)
   (d) Physiological disorder

5. **Small leaf and little leaf respectively are**
   (a) Viral disease of cotton, a mycoplasmal disease of brinjal
   (b) Mycoplasmal disease of citrus, a mycoplasmal disease of brinjal
   (c) Both the terms are synonymous and is a mycoplasmal disease of brinjal
   (d) None of the above

6. **Yellow dragon disease is related to**
   (a) A bacteria that stains pink in gram staining and effects citrus
   (b) A bacterium that stains violet in gram staining and effects citrus
   (c) A virus that is rod shaped and effects citrus
   (d) A mycoplasma that is wall less and effects citrus

7. **Greening disease is**
   (a) Viral disease transmitted by *toxoptera citricidus*
   (b) Bacterial disease transmitted by *pentalonia nigronervosa*
   (c) Viral disease transmitted by *diaphorina citri*
   (d) Bacterial disease transmitted by *diaphorina citri*

8. **Choose the correct combination**
   (a) *Toxoptera citricida*- psyllid
   (b) *Pentalonia nigronervosa*- aphid
   (c) *Diaphorina citri*- psyllid
   (d) Both (a) and (b)

9. **Whiptail of cauliflower and whiptail of sugarcane are respectively**
   (a) Both are nutrient deficiencies of mo
   (b) Both are diseases caused by fungi
   (c) Nutrient deficiency of mo, disease caused by fungus
   (d) Nutrient deficiency of mo, disease caused by mycoplasma

10. **Which pathogen infects potatoes in an early stage**
    (a) *Phytophthora infestans* (b) *Alternaria solani*
    (c) *Streptomyces scabies* (d) *Rhizoctonia solani*

11. **Sclerotia on the potato tubers are a sign of**
    (a) Warts (b) Scab
    (c) Black scurf (d) All of the above

12. Sesamum phyllody is transmitted by

(a) Aphids
(b) Psyllids
(c) Leaf hoppers
(d) Plant hoppers

13. Pineapple disease is

(a) Caused by si deficiency in pineapple
(b) Caused by fungus in pineapple
(c) Caused by si deficiency in sugarcane
(d) Caused by fungus in sugarcane

14. Identify the correct combination

(a) Red stripe- *colletotrichum falcatum*
(b) Red rot- *xanthomonas rubrilineans*
(c) Both (a) and (b)
(d) None of the above

15. Culmicolous smut of sugarcane is caused by

(a) *Claviceps sachhari* and affects the leaves
(b) *Ustilago scitamanea* and affects the leaves
(c) *Ustilago scitamanea* and affects the stalk
(d) *Ustilago sacchari* and affects the stalk

16. Pathogen attacking more severely in ratoon crops than on those planted as setts

(a) Gram positive bacterium clavibacter
(b) Gram negative bacterium clavibacter
(c) Gram negative bacterium xanthomonas
(d) Gram positive bacterium xanthomonas

17. Pink discolouration of vascular bundles in pith is a symptom of

(a) Red rot of sugarcane
(b) Ratoon stunting disease
(c) Gummosis
(d) Both (a) and (b)

18. Characteristic alcoholic smell is seen in sugarcane caused by

(a) *Colletotrichum falcatum*
(b) *Ustilago scitamanea*
(c) *Clavibacter xyli*
(d) *Xanthomonas axonopodis*

19. White pustules in mustard are caused due to

(a) *Peronospora brassicae* whose cell wall is made up of chitin
(b) *Albugo candida* whose cell wall is made up of chitin

(c) *Albugo candida* whose cell wall is made up of cellulose

(d) *Peronospora brassicae* whose cell wall is made up of cellulose

20. Collar rot in groundnut is caused by

(a) *Apergillus niger* (b) *Aspergillus flavus*

(c) Both (a) and (b) (d) None of the above

21. Green plague is associated with

(a) Viral disease of green gram

(b) Bacterial disease of green gram

(c) Viral disease of red gram

(d) Bacterial disease of red gram

22. Cottony growth in barley is due to

(a) *Ustilago hordei* (b) *Ustilago nuda*

(c) *Erysiphe graminis* (d) Both (a) and (b)

23. *Claviceps fusiformis* is

(a) Seed borne (b) Soil borne

(c) Both (a) and (b) (d) None of the above

24. Spindle shaped spots on rice is a symptom of

(a) Sheath blight (b) Brown spot

(c) Blast (d) Tundu

25. Constituents of rice tungro virus

(a) Rtsv- dna virus, rtbv- rna virus

(b) Rtsv- rna virus, rtbv- dna virus

(c) Rtsv and rtbv are rna both

(d) Rtsv and rtbv are dna both

26. Brown spot of rice is more severe in soils like

(a) N deficient soils (b) P deficient soils

(c) K deficient soils (d) Zn deficient soils

27. Cancer of wheat is

(a) Karnal bunt (b) Partial bunt

(c) Complete bunt (d) Both (a) and (b)

28. Leaf rust of wheat is caused by

(a) *Puccinia striformis* (b) *Puccinia graminis*

(c) *Puccinia recondita* (d) None of the above

29. Killer disease of wheat is

(a) Black rust
(b) Brown rust
(c) Yellow rust
(d) All of the above

30. Earliest appearance of rust on wheat in India

(a) Leaf rust
(b) Stem rust
(c) Stripe rust
(d) All of the above

31. Damage of this rust is less when compared to other rusts

(a) Stem rust
(b) Stripe rust
(c) Leaf rust
(d) None of the above

32. Khaira disease is due to deficiency of

(a) Mo
(b) N
(c) Zn
(d) P

33. Offensive fishy odour of grains effected by karnal bunt is due to

(a) Methyl amine
(b) Di methyl amine
(c) Tri methyl amine
(d) Both (a) and (b)

34. Foot rot of rice is also known as

(a) Bakane disease
(b) Foolish seedling disease
(c) Both (a) and (b)
(d) Sehu

35. Udbatta disease of rice is caused by

(a) *Rhizoctonia solani*
(b) *Claviceps oryzae*
(c) *Helminthosporium oryzae*
(d) *Ephelis oryzae*

36. Draining of excess water is done for

(a) Rice blast
(b) Tungro disease
(c) Pansukh
(d) Bakane disease

37. Green ear disease of bajra is

(a) Powdery mildew
(b) Downy mildew
(c) Phytoplasmal disease
(d) Viral disease

38. Yellowish slimy ooze is caused by

(a) *Clavibacter tritici*
(b) *Anguina tritici*
(c) Both (a) and (b)
(d) None of the above

39. Flotation of seeds in brine solution is an effective measure for

(a) Tundu
(b) Loose smut
(c) Flag smut
(d) Karnal bunt

40. Loose smut of wheat pathogen is

(a) Internally seed borne
(b) Externally seed borne
(c) Soil borne
(d) Air borne

41. Karnal bunt is

(a) Seed borne
(b) Soil borne
(c) Air borne
(d) All of the above

42. Sugary disease of sorghum is caused by

(a) *Claviceps fusiformis*
(b) *Sphacelia sorghi*
(c) *Puccinia purpurea*
(d) *Sphecelotheca reiliana*

43. Apical portion of leaf becomes white in maize due to

(a) Downy mildew
(b) Powdery mildew
(c) White bud
(d) Charcoal rot

44. Leaf curl of tomato is transmitted by

(a) Aphids
(b) Whiteflies
(c) Psyllids
(d) Leaf hoppers

45. Capsule rot of cardomum is caused by

(a) Pythium
(b) Phytophthora
(c) Xanthomonas
(d) Both (a) and (b)

46. Blossom end rot of tomato is caused by deficiency of

(a) Zn
(b) Ca
(c) B
(d) K

47. Black heart of potato is due to

(a) O_2 deficiency
(b) High temperature
(c) Both (a) and (b)
(d) Streptomyces scabies

48. Water-soaked lesions on brinjal is due to

(a) *Altenaria solani*
(b) *Colletotrichum capsici*
(c) *Phomopsis vexans*
(d) Phytoplasma

49. Black arm of cotton is due to

(a) O_2 deficiency
(b) Bacterial blight
(c) Angular leaf spot
(d) Both (b) and (c)

50. Club shaped roots are formed due to

(a) Mo deficiency
(b) *Plasmodiophora brassicae*
(c) B deficiency
(d) *Albugo candida*

ANSWER KEY

1	2	3	4	5	6	7	8	9	10	11	12	13	14	15
(c)	(d)	(d)	(d)	(a)	(a)	(d)	(d)	(c)	(b)	(c)	(c)	(d)	(d)	(c)
16	17	18	19	20	21	22	23	24	25	26	27	28	29	30
(a)	(b)	(a)	(c)	(a)	(c)	(c)	(c)	(c)	(b)	(c)	(d)	(c)	(a)	(a)
31	32	33	34	35	36	37	38	39	40	41	42	43	44	45
(a)	(c)	(c)	(c)	(d)	(c)	(b)	(c)	(a)	(a)	(d)	(b)	(c)	(b)	(b)
46	47	48	49	50										
(b)	(c)	(c)	(d)	(b)										

HORTICULTURE

Chapter 20

Horticulture Important Facts, Institutions (Horticultural), Fruit Type, Edible Part, Nutrition Value of Fruits, Major Nutritional Deficiency Diseases, Causes of Colour, Bitterness, Epithets for Fruits, Training and Pruning and their Methods

1. Green colour in potato is due to

(a) Potatin (b) Solanin

(c) Tuberosin (d) Igrin

2. Ca content is high in one of the following vegetables

(a) Bael leaves (b) Tapioca

(c) Agathi (d) Potato

3. Daily requirement of vegetable

(a) 150 g (b) 200 g

(c) 300g (d) 80 g

4. National research center for citrus located at

(a) Lucknow (b) Surat

(c) Pune (d) Nagpur

5. Which of the following countries has highest fruit production in the world

(a) America (b) India

(c) Russia (d) China

6. **Pick the wrong one**
   (a) Adam's fig-Banana
   (b) King of forest-teak
   (c) Kalpavriksha-Arecanut
   (d) Poor man's fruit-Ber

7. **Bael has highest quantity of**
   (a) Vit A
   (b) Ca
   (c) Vit B5
   (d) Riboflavin

8. **Horticulture derived from**
   (a) French
   (b) Arabic
   (c) Greek
   (d) Latin

9. **National horticulture board established in the year**
   (a) 1984
   (b) 1967
   (c) 2005
   (d) 2014

10. **Which of the following training system is known as 6 cane system**
   (a) Head system
   (b) Kniffin system
   (c) Telephone system
   (d) Pandal system

11. **Type of pruning that is generally practiced in fig**
   (a) Ringing
   (b) Girdling
   (c) Thinning
   (d) Notching

12. **Which of the following plants are used for Topiary**
   (a) Duranta
   (b) Sunrose
   (c) Alternanthera
   (d) All of the above

13. **Ringing is generally practiced in**
   (a) Fig
   (b) Sunrose
   (c) Grape
   (d) Guava

14. **Bonsai is a**
   (a) Malaysian art of growing trees
   (b) Japanese art of growing trees
   (c) Chinese art of growing trees
   (d) Swiss art of growing trees

15. **Buttoning of cauliflower is due to def of**
   (a) Mo
   (b) N
   (c) Fe
   (d) Mg

16. As of 2022 which country is top steel producer in the world

(a) India
(b) Russia
(c) China
(d) America

17. Which country launched the Hwasong-12 mid-range ballistic missile?

(a) India
(b) South Korea
(c) Russia
(d) North Korea

18. Which institution released the 'Death Penalty in India' Report?

(a) Supreme court
(b) National Law University, Delhi
(c) NITI Aayog
(d) National cyber-crime bureau

19. Trade emerge is an online platform of which bank

(a) SBI
(b) Union bank
(c) ICICI
(d) Canara Bank

20. Who has been appointed as the Chairman of India Tourism Development Corporation (ITDC)?

(a) Subramanian Swamy
(b) Lal Singh Arya
(c) Sambit Patra
(d) Jamal Siddiqui

21. First country in the world to adopt a ban of deforestation

(a) America
(b) Scotland
(c) Switzerland
(d) Norway

22. World cotton day celebrated on

(a) Oct 7th
(b) Mar 6th
(c) Sep 16
(d) Apr 2nd

23. India's first pollinator park located at

(a) UK
(b) Uttar Pradesh
(c) Andhra Pradesh
(d) Gujarat

24. First state in the country to flagged of first fruit train is

(a) Andhra Pradesh
(b) Uttar Pradesh
(c) Madhya Pradesh
(d) Telangana

25. India's first honey testing Laboratory set up at

(a) NDDB Gujarat
(b) NBAIR Bangalore
(c) IARI New Delhi
(d) NDRI Karnal

26. Edible part of strawberry is

(a) Placenta (b) Fleshy aril

(c) Thalamus (d) Pericarp

27. Carrot is a

(a) Balausta (b) Nut

(c) Capsule (d) Schizocarp

28. Berry is also known as

(a) Stone (b) Asthil

(c) Balausta (d) Saras

29. Fruit cracking in Apple is due to

(a) K (b) B

(c) P (d) N

30. Grey spike of oat is due to deficiency of

(a) Mn (b) Zn

(c) Fe (d) Pb

31. Whip tail of cauliflower is due to deficiency of

(a) Mo (b) Mn

(c) Fe (d) Ca

32. One among the following is century plant

(a) Coconut (b) Arecanut

(c) Datepalm (d) Banana

33. Which of the following fruit has more iron

(a) Walnut (b) Dry karonda

(c) Fig (d) Guava

34. Which of the following vegetable has high calorific value

(a) Walnut (b) Almond

(c) Amaranthus (d) Tapioca

35. Orange colour in mango is due to

(a) Alpha carotene (b) Beta carotene

(c) Gamma carotene (d) Delta carotene

36. Pungency in Raphanus is due to

(a) Glucosinolates (b) Isothiocynates

(c) Oleoresins (d) Allyl di sulphides

37. Which of the following training system of grape is most adopted in India

(a) Head system (b) Telephone system
(c) Kniffin system (d) Pendal system

38. Red colour in carrot is due to

(a) Lycopene (b) Quercetin
(c) Anthocyanins (d) Carotene

39. Cocoa is also known as

(a) White gold (b) Food of god
(c) King of arid fruits (d) Pseudofig

40. National horticulture mission started in the year

(a) 2001 (b) 1997
(c) 2016 (d) 2005

41. Central institute for subtropical horticulture located at

(a) Srinagar (b) Lucknow
(c) Calicut (d) Hyderabad

42. Central research institute for chikoo is located at

(a) Muzaffarpur (b) Puttur
(c) Tiruchirappalli (d) Cochin

43. Largest producer of kiwi fruit in India

(a) Shimla (b) Assam
(c) Manipur (d) Himachal Pradesh

44. Highest production of vegetables in the state

(a) West Bengal (b) Andhra Pradesh
(c) Uttar Pradesh (d) Madhya Pradesh

45. King of fruits

(a) Ber (b) Banana
(c) Apple (d) Mango

46. Which of the following fruit is known as Miracle fruit

(a) Cocoa (b) Gladiolus
(c) China kiwi fruit (d) Litchi

47. Study of vegetables is known as

(a) Pomology (b) Arobriculture

(c) Olericulture (d) Floriculture

48. Which of the following is not a drupe

(a) Guava (b) Almond

(c) Coconut (d) Jamun

49. Fleshy receptacle is edible part of

(a) Litchi (b) Pineapple

(c) Fig (d) Grape

50. Bael is a

(a) Nut (b) Amphisarca

(c) Syconus (d) Sorosis

ANSWER KEY

1	2	3	4	5	6	7	8	9	10	11	12	13	14	15
(b)	(c)	(c)	(d)	(d)	(c)	(d)	(d)	(a)	(c)	(d)	(a)	(c)	(b)	(b)
16	17	18	19	20	21	22	23	24	25	26	27	28	29	30
(c)	(d)	(b)	(c)	(c)	(d)	(a)	(a)	(a)	(a)	(c)	(d)	(d)	(b)	(a)
31	32	33	34	35	36	37	38	39	40	41	42	43	44	45
(a)	(c)	(b)	(d)	(a)	(b)	(d)	(c)	(b)	(d)	(b)	(b)	(c)	(a)	(d)
46	47	48	49	50										
(c)	(c)	(a)	(c)	(b)										

Chapter 21

Orchard Layout, Commercial Propagation Methods, Families of Fruit Crops, Fruit Production of Mango, Banana, Citrus, Papaya, Pomegranate, Ber, Guava, Amla, Grape and Coconut

1. **Leading country/state in fruit production**
   (a) China (b) India
   (c) Kashmir (d) West Bengal

2. **Most acceptable training system acceptable**
   (a) Head system (b) Kniffin
   (c) Telephone (d) Modified leader system

3. **Land slope required for orchard should not be more than**
   (a) 3 per cent (b) 5 per cent
   (c) 10 per cent (d) 15 per cent

4. **Training is related to**
   (a) To make plant strong to bear fruit load
   (b) To obtain good quality yield
   (c) To give proper shape
   (d) Both (a) and (b)

5. **Which one is wrongly related**
   (a) Bonsai- short stature (b) Topiary- shape
   (c) Pruning-qualitative yield (d) Hedge-perennial plant

6. Notching is done in which plant

(a) Mango (b) Grape

(c) Fig (d) Litchi

7. Ringing is followed in

(a) Grape (b) Ring tree

(c) pomegranate (d) Karonda

8. Number of plants can be obtained through formula

(a) 10000/RXR (b) RXR/10000

(c) (10000/RXR) * 100 (d) (RXR/10000) *100

9. Simplest and widely adopted method of planting

(a) Square (b) Quincux

(c) Hexagonal (d) Rectangle

10. In which method right angle planting is followed

(a) Square (b) rectangle

(c) Quincux (d) Hexagonal

11. Which one is wrongly related

(a) Hexagonal-15 per cent more than square

(b) Quincux-double than square

(c) Square- twice the product of spacing

(d) Countour- Hilly areas

12. Meadow system is originated in

(a) India (b) Iran

(c) Iraq (d) Israel

13. Amrapali is cross between

(a) Dasheri x Neelum (b) Neelum x Dasheri

(c) Banganapalli x arka aruna (d) Totapuri x Niranjan

14. Which system used in high density planting

(a) Triangular system (b) Quincux system

(c) Contour system (d) Equilateral triangle system

15. which one is wrongly matched

(a) Tea- soft wood cutting (b) Grape- softwood

(c) Mango- epicotyl (d) Banana- sword sucker

16. **In which crop stooling and inarching is considered standard for propagation respectively**

(a) Rose, Olive (b) Ber, bael

(c) Sapota, Guava (d) Guava, sapota

17. **Corm is propagative material of**

(a) Rose (b) Gladiolous

(c) Tuber (d) Lily

18. **Which among the following is not a day neutral plant**

(a) Papaya (b) Guava

(c) Banana (d) pineapple

19. **Fruit type wrongly related**

(a) Balusta-Pomegranate (b) Amphisarca- Bael

(c) Synconus- fig (d) Schizocarp- radish

20. **Edible part of strawberry**

(a) Thalamus (b) Endocarp

(c) Seeds (d) Pericarp

21. **Salinity tolerant fruit crop**

(a) Datepalm (b) Kair

(c) Ber (d) All of the above

22. **Pollination in mango aided by**

(a) Root grub (b) mealy bug

(c) House fly (d) Whitefly

23. **Vapour hear treatment in mango done against which pest**

(a) Fruitfly (b) Stone weevil

(c) Mealy bug (d) Mango hopper

24. **Highest and lowest bisexual flowers are seen in varieties**

(a) Saisuganda and sai sarbhati (b) Sai - suganda and Rumani

(c) Alphonso and Sindhu (d) Niranjan and Fazli

25. **Variety suitable for long transport**

(a) Lal sindhuri (b) Sagar

(c) Neelum (d) Banglura

26. Quarantine important pest in mango

(a) Mealy bug
(b) hopper
(c) Whitefly
(d) Fruitfly

27. Diploid variety of banana

(a) Okra
(b) Lady finger
(c) Sunu
(d) Cavendish

28. Banana storage temperature

(a) 10°C
(b) 13°C
(c) 20°C
(d) 37°C

29. Red skin banana

(a) Poovan
(b) Red love
(c) Valentine
(d) Lal vachi

30. Moko disease of banana is caused by

(a) Virus
(b) Bacteria
(c) Fungus
(d) Phytoplasma

31. Kinnow is

(a) Mandarin
(b) Sweet orange
(c) Lime
(d) Yellow lime

32. Citrus family is

(a) Musaceae
(b) Rutaceae
(c) Annonaceae
(d) Caricaceae

33. Granulation in lime is due to

(a) High temperature
(b) Low temperature
(c) High humidity
(d) Freezing

34. Which term wrongly relate to fruit crop papaya

(a) Pensin
(b) Caricaxanthin
(c) Papain
(d) Allicin

35. Pusa nanha is variety of papaya developed through

(a) Hybridization
(b) Backcross breeding
(c) Mutation
(d) Micro propagation

36. Coorg's is a variety of

(a) Rubber
(b) Coffee
(c) Tea
(d) Lime

37. Pomegranate is good fruit for patients suffering from

(a) Blood cancer
(b) Leprosy
(c) Luekemia
(d) Genetic disorder

38. Ruby and Pusa ruby are varieties of _______ respectively

(a) Pomegranate and Tomato
(b) Cabbage and cauliflower
(c) Tomato and apple
(d) Apple and tomato

39. Anar butterfly

(a) Virachola isocrates
(b) Deudorix isocrates
(c) Both
(d) Eudrillus euginae

40. Which acid is present in grape

(a) Benzoic acid
(b) Tartaric acid
(c) Ally tartaric acid
(d) Synzenoic acid

41. Apple bowl of India

(a) Kashmir
(b) Dehradun
(c) Himachal pradesh
(d) Delhi

42. ICMR Reccomendation for fruits and vegetables

(a) 100gm and 300g/person/day
(b) 150g and 350g/person/day
(c) 50g and 150g/person/day
(d) 580g and 120g/person/day

43. Horticulture production increased to how percent in 2020-2021 from previous 2019-2020.

(a) 2.93 per cent
(b) 2.32 per cent
(c) 5.23 per cent
(d) 10 per cent

44. Top producer of vegetables state

(a) Karnataka
(b) Odisha
(c) West bengal
(d) Rajasthan

45. National horticulture board located at

(a) Gurugram, haryana
(b) Hesaraghatta, Bangalore
(c) Mahbubnagar, Telangana
(d) Theni, Tamil Nadu

46. Which state framed a 'draft policy for senior citizens' recognising demographic transition?

(a) Karnataka
(b) Tamil Nadu
(c) Andhra Pradesh
(d) Kerala

47. Which state has approved EX-gratia relief of Rs 20000 per acre to farmers?

(a) West Bengal
(b) Delhi
(c) Punjab
(d) Uttar Pradesh

48. India approved the 'digital work plan 2022' along with which bloc?

(a) SAARC
(b) ASEAN
(c) G-20
(d) European commission

49. India's first geo-park in which state?

(a) Maharashtra
(b) Telangana
(c) Karnataka
(d) Madhya Pradesh

50. Masada Forest, which was seen in the news is located in which country?

(a) India
(b) Sri Lanka
(c) Israel
(d) UAE

ANSWER KEY

1	2	3	4	5	6	7	8	9	10	11	12	13	14	15
(a)	(d)	(b)	(d)	(d)	(c)	(a)	(a)	(a)	(a)	(c)	(d)	(a)	(a)	(b)
16	17	18	19	20	21	22	23	24	25	26	27	28	29	30
(d)	(b)	(d)	(d)	(a)	(d)	(c)	(a)	(b)	(c)	(d)	(b)	(b)	(d)	(b)
31	32	33	34	35	36	37	38	39	40	41	42	43	44	45
(a)	(b)	(a)	(d)	(c)	(b)	(b)	(a)	(c)	(b)	(c)	(a)	(a)	(c)	(a)
46	47	48	49	50										
(b)	(b)	(b)	(d)	(c)										

Chapter 22

Vegetable Production, Classification of Vegetables, Production of Tomato, Brinjal, Chilli, Okra, Cole Crops, Cabbage, Cauliflower, Onion, Radish, Carrot, Cucurbits, Cucumber, Watermelon, Muskmelon, Bottle Gourd, Pea, Cowpea, Cluster Bean, Palak

1. **World's largest producer of vegetable**
   (a) India (b) China
   (c) US (d) Turkey
2. **Which method of vegetable cultivation is extensive?**
   (a) Truck gardening (b) Kitchen gardening
   (c) Market gardening (d) Roof gardening
3. **Botanical name of ridge gourd**
   (a) *Luffa acutangula* (b) *Luffa cylindrica*
   (c) *Lagenaria siceraria* (d) *Sauropus androgynus*
4. **A dioecious annual plant**
   (a) Ivy gourd (b) Pointed gourd
   (c) Round melon (d) Spinach
5. **Arka Manik is a variety of**
   (a) Bottle gourd (b) Cucumber
   (c) Pumpkin (d) Watermelon

6. **Pusa Manjari and Pusa meghdoot are varieties of**
   (a) Ridge gourd (b) Bottle gourd
   (c) Sponge gourd (d) French bean

7. **Mutant variety of French bean**
   (a) Pusa nasdar (b) Pusa Parvati
   (c) Pusa chikni (d) Pusa sanyog

8. **Family of cassava**
   (a) Umbelliferae (b) Solanaceae
   (c) Euphorbiaceae (d) Malvaceae

9. **Family of lettuce**
   (a) Chenopodiaceae (b) Amaranthaceae
   (c) Apiaceae (d) Asteraceae

10. **Cowpea is probably a native of**
    (a) China (b) Africa
    (c) India (d) None of the above

11. **An early variety of cauliflower is**
    (a) Pusa katki (b) Snowball-16
    (c) Both (a) and (b) (d) Synthetic second

12. **Non climacteric vegetable**
    (a) Tomato (b) Watermelon
    (c) Muskmelon (d) Cucumber

13. **Pungency in garlic is due to**
    (a) Allyl isothiocyanate (b) Allyl propyl disulphide
    (c) Diallyl disulphide (d) Dimethyl disulphide

14. **Nature of pollination in onion**
    (a) Insect (b) Water
    (c) Wind (d) Birds

15. **Fruit type of cabbage**
    (a) Head (b) Siliqua
    (c) Curd (d) All of the above

16. **Which place in Maharashtra is the biggest onion market in India?**
    (a) Lonavala (b) Lasalgaon
    (c) Nashik (d) Pimpalgaon

17. Most popular canned vegetable in India

(a) Tomato (b) Cucumber

(c) Pea (d) Green beans

18. Vegetable known as plant insulin

(a) *Sechium edule* (b) *Luffa cylindrica*

(c) *Benincasa hispida* (d) *Momordica charantia*

19. A beverage prepared from black carrot

(a) Kanji (b) Sauerkraut

(c) Feni (d) Kefir

20. A dessert vegetable is

(a) *Momordica charantia* (b) *Citrullus lanatus*

(c) *Allium cepa* (d) *Capsicum annum*

21. *Trichosanthes dioica* has which of the following flower colours?

(a) Yellow (b) Red

(c) White (d) Orange to yellow

22. Turnip belongs to the family

(a) Umbelliferae (b) Solanaceae

(c) Euphorbiaceae (d) Cruciferae

23. Tomato fruit cracking is caused by

(a) Insufficient watering

(b) A period of drought followed by sudden heavy irrigation

(c) Excessive watering

(d) All of the above

24. Phomopsis blight resistant variety of brinjal

(a) Azad Kranti (b) Pusa Komal

(c) Pusa Bhairav (d) Arka Keshav

25. Day neutral and spineless variety of bhindi

(a) Pusa Sawani (b) Pusa mukhmali

(c) Parbhani Kranti (d) Arka Anamika

26. "Durgapura Kesar" is an important variety of

(a) Cucumber (b) Round melon

(c) Muskmelon (d) Watermelon

27. Which is not correct?

(a) Pusa Gaurav - Tomato
(b) Pusa Rituraj - Brinjal
(c) Pusa Jwala - Chilli
(d) Pusa Mukta – cauliflower

28. A determinate variety of tomato

(a) Punjab chhuhara
(b) Roma
(c) Parker
(d) Pusa-120

29. A serious disorder of tomato grown in greenhouses or protected structures

(a) Fruit cracking
(b) Blossom end rot
(c) Puffiness
(d) Sun scald

30. A physiological disorder in carrot

(a) Pithiness
(b) Forking
(c) Buttoning
(d) Hollow heart

31. Water soluble amino acid present in garlic

(a) Allicin
(b) Allin
(c) Both (a) and (b)
(d) None of the above

32. Grain purpose cluster bean variety is

(a) Pusa sadabahar
(b) Pusa mausmi
(c) Pusa navbahar
(d) Durgapura safed

33. Pea maturity is measured by

(a) Arealometer
(b) Peatometer
(c) Tendrometer
(d) Hydrometer

34. Most serious disease of Bhindi

(a) Yellow vein mosaic
(b) Powdery mildew
(c) Root rot
(d) None of the above

35. Browning in cauliflower is due to deficiency of

(a) Nitrogen
(b) Boron
(c) Molybdenum
(d) Magnesium

36. Chemical used for controlling sprouting of onion during storage

(a) Ethylene
(b) Ethion
(c) Maleic hydrazide
(d) All of the above

37. Variety of chilli suitable for high density planting

(a) Jwalamukhi (b) Jwalasakhi

(c) Pusa Jwala (d) Punjab lal

38. Which is a popular variety of pumpkin?

(a) Pusa nasdar (b) Pusa visesh

(c) Pusa vishwas (d) Pusa lal

39. Photo periodically, potato is

(a) Long day plant (b) Short day plant

(c) Day neutral (d) None of the above

40. Anticancer property of cabbage is due to

(a) Dimethyl trisulphite (b) Quecertin

(c) Indole-3- carbinol (d) Solanine

41. Removal of central portion of curd for earlier initiation of flower stalk in cauliflower is called

(a) Blanching (b) Bolting

(c) Stackling (d) Scooping

42. Which state has introduced 'Night safaris' in its national parks?

(a) Odisha (b) Madhya Pradesh

(c) Maharashtra (d) Punjab

43. This year, the theme of World Pulses Day is "Pulses to empower youth in achieving sustainable agri-food systems". World Pulses Day is celebrated on

(a) January 29 (b) February 2

(c) February 5 (d) February 10

44. Which Board/Organization has launched the National Nursery Portal?

(a) National Housing Board (b) National Horticulture Board

(c) NITI Aayog (d) National Biodiversity Board

45. Asia's first rice technology park established at

(a) Karnataka (b) Telangana

(c) Odisha (d) Punjab

46. The Jal Jeevan Mission is a flagship programme of union govt. which aims to provide tap water connection to every rural household by which year?

(a) 2022 (b) 2023

(c) 2024 (d) 2025

47. Indian agriculture is typically characterized as

(a) Land surplus and labour scarce economy

(b) Land surplus and labour surplus economy

(c) Land scarce and labour surplus economy

(d) Land scarce and labour scarce economy

48. The headquarters of World Food prize foundation is at

(a) De Moines (USA) (b) Rome (Italy)

(c) Manila (Philippines) (d) Washington (USA)

49. India's first food museum launched in which state?

(a) Kerala (b) Telangana

(c) Andhra Pradesh (d) Tamil Nadu

50. Mission of Ramsar convention is to conserve

(a) Wetlands (b) Rivers

(c) Oceans (d) Deserts

ANSWER KEY

1	2	3	4	5	6	7	8	9	10	11	12	13	14	15
(b)	(a)	(a)	(d)	(d)	(b)	(b)	(c)	(d)	(b)	(a)	(d)	(c)	(a)	(b)
16	17	18	19	20	21	22	23	24	25	26	27	28	29	30
(b)	(c)	(d)	(a)	(b)	(c)	(d)	(b)	(c)	(a)	(d)	(d)	(a)	(b)	(b)
31	32	33	34	35	36	37	38	39	40	41	42	43	44	45
(b)	(d)	(c)	(a)	(b)	(c)	(a)	(c)	(a)	(c)	(d)	(b)	(d)	(b)	(a)
46	47	48	49	50										
(c)	(c)	(a)	(d)	(a)										

Chapter 23

Floriculture, Famous Gardens and Style of Gardens, Features of Garden, Flower Crops like Rose, Gladiolus, Chrysanthemum, Mary Gold, Fruit Preservation Principles, Jam, Jelly, Squash, Cordial, Sharbat, Preserve, Candy, Bonsai, Topiary, Hedges, Edge

1. **Kyoto protocol was conducted on**
   (a) December 1997 (b) December 1998
   (c) January 1997 (d) January 1998
2. **Kigali agreement was launched**
   (a) 2015 (b) 2016
   (c) 2017 (d) 2018
3. **First agricultural commodity to enter agricultural commodities market**
   (a) Cluster bean (b) Chilli
   (c) Maize (d) Both (a) and (b)
4. **GST on vegetables**
   (a) 0 per cent (b) 5 per cent
   (c) 8 per cent (d) 10 per cent
5. **South India's first organic farming university may set up in**
   (a) Karnataka (b) Kerala
   (c) Andhra Pradesh (d) Telangana

6. **Best example of Japanese art in growing miniature tree is done for which tree sp.**
   (a) *Cassia* sp.
   (b) *Eucalyptus* sp.
   (c) *Ficus* sp.
   (d) *Clerodendron* sp.

7. **Fencing is a function of**
   (a) Edge
   (b) Hedge
   (c) Topiary
   (d) Bonsai

8. **Killing of both beneficial and harmful microbes is done in**
   (a) Sterilization
   (b) Pasteurization
   (c) Canning
   (d) Fermentation

9. **Plan of gardening is symmetrical in**
   (a) Informal style
   (b) Formal style
   (c) Both of these
   (d) Free style

10. **Thornless variety of rose is**
    (a) Suchitra
    (b) Pusa mohit
    (c) Tushar
    (d) Both (a) and (b)

11. **Cryopreservation is preservation in**
    (a) Liquid nitrogen at 196 C
    (b) Liquid nitrogen at - 196 C
    (c) Liquid nitrogen at 96 C
    (d) Liquid nitrogen at - 96 C

12. **Fruit acid content in Jelly is**
    (a) 1 per cent
    (b) 1.5 per cent
    (c) 0.5 per cent
    (d) 0.25 per cent

13. **Acid content in Jam is**
    (a) 0.5 per cent
    (b) 1 per cent
    (c) 2 per cent
    (d) 3 per cent

14. **Japanese garden is an example of**
    (a) Asymmetrical type
    (b) Symmetrical type
    (c) Both formal and informal
    (d) None of the above

15. **Maximum area under floriculture is**
    (a) Telangana
    (b) Kerala
    (c) Maharashtra
    (d) Karnataka

16. **African marigold is**
(a) Large flowered *T. erecta*
(b) Small flowered *T. erecta*
(c) Large flowered *T. patula*
(d) Small flowered *T. patula*

17. **Gulkand is made by mixing**
(a) Bracts + Petals + sugar in 1:1:1 ratio
(b) Petals + Sugar in 1:1 ratio
(c) Petals + Sugar in 1:2 ratio
(d) Petals + Sugar in 1.5:2 ratio

18. **Guldaudi is a common name of**
(a) Gladiolus
(b) Rose
(c) Chrysanthemum
(d) Marigold

19. **Gladiolus belongs to**
(a) Asteracae
(b) Iridacae
(c) Compositae
(d) None of the above

20. **National botanical garden is located in**
(a) Uttar Pradesh
(b) Uttarakhand
(c) Andhra Pradesh
(d) Arunachal Pradesh

21. **Which flower has maximum cultivation**
(a) Rose
(b) Jasmine
(c) Marigold
(d) Chrysanthemum

22. **Biggest Garden having symmetrical style of gardening is**
(a) Mughal garden
(b) Persian garden
(c) Pinjore garden
(d) Brindavan garden

23. **Biggest Garden having symmetrical style of garden is located in**
(a) Karnataka
(b) Kerala
(c) Telangana
(d) Goa

24. **Garden representing natural beauty**
(a) Mughal
(b) English
(c) Japanese
(d) Persian

25. **To increase the storage period of flowers, they are subjected to**
(a) Chilling
(b) Cryopreservation
(c) Pulsing
(d) Blanching

26. Chemical used for colourless fruit juices and the actual compound responsible for it

(a) KMS, SO_2
(b) KMS, K_2O
(c) Sodium benzoate, Benzoic acid
(d) Sodium benzoate, Sodium

27. which of the following does not contain fruit juice

(a) Sarbat
(b) Squash
(c) Cordial
(d) All of the above

28. Agra petha is a candy made of

(a) Amla
(b) Ash gourd
(c) Ginger
(d) Squash

29. Mandhor garden is located in

(a) Blue city
(b) Pink city
(c) Green city
(d) None of the above

30. Rockery is an important component of

(a) Japanese garden
(b) Persian garden
(c) English garden
(d) Rose garden

31. Unavoidable and most important component of a garden is

(a) Rockery
(b) Lawn
(c) Baradari
(d) Garden lanterns

32. National flower of England is originated from

(a) England
(b) India
(c) USA
(d) Australia

33. Baradari is a component of

(a) English
(b) Persian
(c) Japanese
(d) Mughal

34. Rose Garden has

(a) Symmetrical type
(b) Asymmetric type
(c) Both (a) and (b)
(d) None of the above

35. Propagation of Guldaudi is through

(a) T budding
(b) Corms
(c) Root suckers
(d) Tissue culture

36. Tushar is a

(a) Red variety of rose
(b) Yellow variety of rose
(c) White variety of rose
(d) Pink variety of rose

37. Temperature and time of pasteurization are

(a) 30°C, 90 minutes
(b) 90°C, 30 minutes
(c) 60°C, 20 minutes
(d) 100°C, 60 minutes

38. Sugar content of jam is

(a) 68 per cent and added according to protein content
(b) 68 per cent and added according to acid content
(c) 68 per cent and added according to both protein and acid content
(d) 100 per cent and added according to both

39. Blanching is important for

(a) Checking enzymatic spoilage in fruits
(b) Checking enzymatic spoilage in vegetables
(c) Checking enzymatic spoilage in both fruits and vegetables
(d) None of the above

40. concentration of salt to act as a preservative

(a) 15 - 25 per cent
(b) 30 - 40 per cent
(c) 50 per cent
(d) 60 per cent

41. Following are true in the context of canning

(a) Processing temperature of 100 C for fruits
(b) Processing temperature of 121 C for vegetables
(c) Processing temperature of 115 C for fruits
(d) Both (a) and (b)

42. Best antioxidant is

(a) Vitamin C
(b) Vitamin b12
(c) Vitamin D
(d) Vitamin A

43. Beautiful gardens is written by

(a) M. S. Swaminathan
(b) M. S. Randhawa
(c) G. S. Khush
(d) G. S. Bhandari

44. Alcohol content in wine

(a) 1 to 5 per cent
(b) 7 to 20 per cent
(c) 20 to 30 per cent
(d) 30 to 40 per cent

45. Concentration of KMS and Sodium Benzoate

(a) 0.05 per cent, 0.07 per cent
(b) 0.07 per cent, 0.05 per cent
(c) 0.5 per cent, 0.7 per cent
(d) 0.7 per cent, 0.5 per cent

46. Which state has framed a draft policy for senior citizens recognizing demographic transition

(a) Tamil Nadu
(b) Kerala
(c) Telangana
(d) Karnataka

47. Who is the first male tennis star to win 21 grand slam titles?

(a) Rafael Nadal
(b) Roger Federer
(c) Novak Djokovic
(d) Andre Agassi

48. United for Dignity is related to

(a) Covid
(b) Leprosy
(c) Cancer
(d) Tuberculosis

49. Which country has developed artificial intelligence Nanny?

(a) China
(b) USA
(c) UK
(d) Germany

50. Cyornis and Zosterops are recently discovered genus of which species from Borneo?

(a) Bird
(b) Beetle
(c) Turtle
(d) Spider

ANSWER KEY

1	2	3	4	5	6	7	8	9	10	11	12	13	14	15
(a)	(b)	(a)	(a)	(a)	(c)	(b)	(a)	(b)	(d)	(b)	(a)	(a)	(a)	(d)
16	17	18	19	20	21	22	23	24	25	26	27	28	29	30
(a)	(b)	(c)	(b)	(a)	(b)	(d)	(a)	(c)	(c)	(a)	(a)	(b)	(a)	(c)
31	32	33	34	35	36	37	38	39	40	41	42	43	44	45
(b)	(b)	(d)	(c)	(c)	(c)	(b)	(b)	(b)	(a)	(d)	(a)	(b)	(b)	(a)
46	47	48	49	50										
(a)	(a)	(b)	(a)	(a)										

ENTOMOLOGY

Chapter 24

Important Points of Entomology, Insect Morphology, Cuticle, Moulting, Types of Larvae, Pupa, Antennae Type, Types of Mouth Parts, Types of Legs, Metamorphosis, Endocrine and Exocrine, Sense Organs, Insect Behaviour

1. **The two types of compound eyes found in insects are**
   (a) Proposition and Superposition
   (b) Apposition and Superposition
   (c) Proposition and Supraposition
   (d) Apposition and Supraposition
2. **"Pyigidium" is the character of**
   (a) Coleoptera (b) Lepidoptera
   (c) Diptera (d) Hymenoptera
3. **Physogastery is character of**
   (a) Butterfly (b) Bug
   (c) Maggots (d) Termites
4. **Inch worms/Earth measures/loopers belong to the family**
   (a) Noctuedae (b) Lymantridae
   (c) Geometridae (d) Galleridae
5. **Pollination by bees is known as**
   (a) Cantharophily (b) Mellitophily
   (c) Myophily (d) Psychophily

6. **Largest and strongest part of leg is**
   (a) Coxa (b) Trochanter
   (c) Tibia (d) Femur

7. **Homomorphism is seen in '**
   (a) Preying mantid (b) Cowbug
   (c) Membracidae (d) Both (b) and (c)

8. **Spur is a**
   (a) Non cellular outgrouth (b) Unicellular outgrowth
   (c) Milticellular out growth (d) None of the above

9. **Proboscis in butterfly modified from**
   (a) Galea of maxillae (b) Lacinia of maxillae
   (c) Both (d) None of the above

10. **The antenna act as a hearing organ in**
    (a) Fleas (b) Ants
    (c) Mosquitoes (d) All of the above

11. **The antenna act as communication in**
    (a) Fleas (b) Ants
    (c) Mosquitoes (d) All of the above

12. **The antenna act as a useful organ to clasp the male**
    (a) Fleas (b) Ants
    (c) Mosquitoes (d) All of the above

13. **The type of antenna where size of the segments decreases from base to apex**
    (a) Setaceous (b) Moniliform
    (c) Filiform (d) Serrate

14. **Antenna enlarges gradually towards the tip**
    (a) Knobbed (b) Clavate
    (c) Clubbed (d) Both (a) and (c)

15. **The terminal segments of antenna become enlarged suddenly**
    (a) Butterfly (b) Blister beetle
    (c) Sawfly (d) Termite

16. **Plumose type of antennae found in**
    (a) Female mosquitoes (b) Male mosquitoes
    (c) Both (d) None of the above

17. **Fan like (flabellate) type antenna found in**
    (a) Strepsipterans (b) Stylopids
    (c) Cedar beetles (d) All of the above

18. **Aristate and geniculate type of antenna found in**
    (a) House fly, mosquitoes (b) Honey bees, house fly
    (c) House fly, honey bee (d) Housefly, thrips

19. **True jaws of insect (primary jaws) are**
    (a) Maxillae (b) Mandibles
    (c) Both (a) and (b) (d) Labium

20. **The accessory jaws of insects (secondary jaws) one**
    (a) Maxillae (b) Mandibles
    (c) Both (a) and (b) (d) Labium

21. **The grosshopper five segmented antenna like maxillary palpi arise from**
    (a) Palpiger (b) Palpifer
    (c) Both (d) None of the above

22. **Right mandible is absent is absent in**
    (a) Thrips (b) Thysanoptera
    (c) Both (a) and (b) (d) None of the above

23. **The glossa in honey bees terminates into a small circular spoon shaped lobe called**
    (a) Spoon (b) Bouton
    (c) Flabellum (d) All of the above

24. **Pseudotrachae in the insect mouth part of**
    (a) Honey bees (b) Housefly
    (c) Mosquitoes (d) All of the above

25. **The foot of insects**
    (a) Tarsus (b) Basitarsus
    (c) Pre tarsus (d) None of the above

26. Two segmented trochanter found in

(a) Dragon fly (b) Damselfly

(c) Ichnemonids (d) All of the above

27. Tympanum organ in mole crickets present at

(a) Basitarsus (b) Pretarsus

(c) Foretibia (d) Hind tibia

28. Corbicula found at

(a) Fore tibia (b) Hind tibia

(c) Fore tarsus (d) Hind tarsus

29. Abdominal legs of caterpillar are called as

(a) Prolegs (b) False legs

(c) Pseudolegs (d) All of the above

30. Pterostigma present in

(a) Dragon fly (b) Damsel fly

(c) Zoraptera (d) Both (a) and (b)

31. The angle by which the wings is attached to the thorax is called

(a) Apical angle (b) Anal angle

(c) Humeral angle (d) Costal angle

32. The anterior area of the wing supported by veins called

(a) Remigium (b) Vannus

(c) Pteralia (d) Jugam

33. Semitransperent spot devoid of hairs in the wings of caddishflies is

(a) Microtrichia

(b) Thyridium with sensory in function

(c) Thyridium with secretory in function

(d) Androconia

34. Grouping of segments into body regions is called

(a) Tagmosis (b) Phragmosis

(c) Segmentation (d) metamorphosis

35. Smoky membranous wing with R thickened is found in

(a) Embioptera (b) Odonata

(c) Orthoptera (d) Grylloblattodea

36. Chitin is abundant in

(a) Exocuticle (b) Cuticullin

(c) Endocuticle (d) Epicutecle

37. Type of wing coupling in mecoptera

(a) Jugate (b) Frenate

(c) Jugo – frenate (d) Hamulate

38. Gula is median ventral plate present in

(a) Isoptera (b) Hymenoptera

(c) Homopetra (d) Heteroptera

39. Visual organ of larval Lepidoptera are

(a) 3 ocelli (b) 3 stemmata

(c) 6 stemmata (d) 6 ommatidia

40. Mouth parts in thrips belong to the type

(a) Chewing (b) Biting

(c) Asymmetrical (d) Sucking

41. Insect head consist of

(a) 5 segments (b) 3 segments

(c) 7 segments (d) 6 segments

42. Technical term of head capsule is

(a) Vertex (b) Epicranial

(c) Cranium (d) Frons

43. The line of weakness is called as

(a) Epistomal suture (b) Frontal suture

(c) Clypeus (d) Epicranial suture

44. Ocelli in cockroach are called as

(a) Ommatidia (b) Fenestrae

(c) Simple eyed (d) None of the above

45. Thorax is made up of how many segments

(a) 4 (b) 3

(c) 6 (d) 5

46. Segmental plate which is present on dorsal side of thorax is called as

(a) Pleuron (b) Sternum

(c) Tergum/Notum (d) Sclerite

47. Lateral plate of thorax is called as

(a) Pleuron (b) Sternum

(c) Notum (d) Sclerite

48. Between the terga and pleuron region which parts are attached to the body

(a) Wings (b) Legs

(c) Cerci (d) Styli

49. Leg consists of primitively how many segments?

(a) 5 (b) 6

(c) 7 (d) 4

50. First tarsal segment is called as

(a) Pre tarsus (b) Basi tarsus

(c) Peri tarsus (d) Epi tarsus

ANSWER KEY

1	2	3	4	5	6	7	8	9	10	11	12	13	14	15
(b)	(a)	(d)	(c)	(b)	(d)	(d)	(c)	(a)	(c)	(b)	(a)	(a)	(b)	(a)
16	17	18	19	20	21	22	23	24	25	26	27	28	29	30
(b)	(d)	(c)	(b)	(a)	(b)	(c)	(d)	(b)	(c)	(d)	(c)	(b)	(d)	(d)
31	32	33	34	35	36	37	38	39	40	41	42	43	44	45
(c)	(a)	(b)	(a)	(a)	(c)	(c)	(d)	(c)	(c)	(d)	(c)	(d)	(b)	(b)
46	47	48	49	50										
(c)	(a)	(b)	(a)	(b)										

Chapter 25

Physiology, Digestive System, Respiratory System, Circulatory System, Excretion, Nervous System

1. **Insect blood sugar is**
   (a) Glucose (b) Trehalose
   (c) Galactose (d) Maltose

2. **Insects reproduce before attaining sexual maturity is known as**
   (a) Paedogenic (b) Embryonic
   (c) Viviparity (d) None of the above

3. **Functional heomaphroditism is a common mode of reproduction in the insect**
   (a) *Platygaster oryzae* (b) *Icerya purchasi*
   (c) Both (a) and (b) (d) None of the above

4. **The vitellogenesis synthesis in fat body is stimulated by ______**
   (a) Pro Thoracico Tropic Hormone (b) Juvenile Hormone
   (c) Ecdysone (d) Allatotropin

5. **Phase polymorphism is shown by ______**
   (a) Grasshoppers (b) Crickets
   (c) Bugs (d) Preying mantis

6. **Plastron respiration is observed in ______**
   (a) Aquatic insects (b) Parasitoids
   (c) Terrestrial insects (d) Aerial insects

7. **Larval body shape in many holometabolous insects such as caterpillars is maintained largely due to**
   (a) Exoskeleton (b) Endoskeleton
   (c) A combination of both the above (d) Hydrostatic pressure
8. **In parthenogenesis where only male is known as**
   (a) Arrhenotoky (b) Thelytoky
   (c) Amphitoky (d) None of the above
9. **Primitive no. of Malphigian Tubules**
   (a) 2 (b) 4
   (c) 6 (d) 10
10. **Excretion and regulation of the internal environment by controlling the ion and water balance in the blood and by removing nitrogenous waste**
    (a) Malphigian Tubules (b) Integument
    (c) Oenocytes (d) None of the above
11. **Principle excretory product in insects**
    (a) Allantoin (b) Honeydew
    (c) Urea (d) Uric acid
12. **______ containing less hydrogen than any other compound of excretion is well adopted for conservation of water**
    (a) Allantoin (b) Honeydew
    (c) Urea (d) Uric acid
13. **In ______ uric acid produced in pupal instar stored in the scales on wings**
    (a) Lemon butterfly (b) Cabbage butterfly
    (c) Pieris brassicae (d) Both (b) and (c)
14. **Basic unit of nervous system**
    (a) Neuron ' (b) Nerve cell
    (c) Both (a) and (b) (d) None of the above
15. **Generally in CNS each body segment will have ______**
    (a) One ganglia (b) Two ganglia
    (c) Three ganglia (d) All of the above

16. The pre oral cavity between epipharynx and hypopharynx is called

(a) Salivarium (b) Cibarium

(c) Both (d) None of the above

17. Pre oral cavity between hypo pharynx and salivary duct is

(a) Salivarium (b) Cibarium

(c) Both (d) None of the above

18. Honey stomach present in

(a) Housefly (b) Honeybees

(c) Both (d) None of the above

19. Mesenteron originated from

(a) Ectoderm (b) Endoderm

(c) Mesoderm (d) All of the above

20. Peritrophic membrane is present in

(a) Sap feeders (b) Solid feeders

(c) Both (d) None of the above

21. Finger like outgrowths found in anterior/posterior ends of midgut are called

(a) Enteric caecae (b) Hepatic caecae

(c) Gastric caecae (d) All of the above

22. Filter chamber is present in

(a) Hemiptera (b) Homoptera

(c) Hymanoptera (d) Lepidoptera

23. Foregut and hind gut originated from

(a) Ectoderm only (b) Ectoderm and endoderm

(c) Mesoderm and ectoderm (d) Endoderm and ectoderm

24. Fermentation chamber is present in

(a) Scarabids (b) Termites

(c) Both (d) None of the above

25. Mid gut opens into hindgut through

(a) Stomodeal valve (b) Pyloric valve

(c) Both (d) None of the above

26. Extra intestinal digestion occurs in

(a) Plant bugs
(b) Ant lion grub
(c) Both
(d) None of the above

27. Enzyme secretion occurs without cell break down called

(a) Holocrine
(b) Merocrine
(c) Both
(d) None of the above

28. Flagellate protozoans present in the gut of

(a) Termites
(b) Wood cockroach
(c) Both
(d) None of the above

29. The structures of Malphigian Tubules are discovered by

(a) M.Malpighi
(b) Stanley
(c) J.Petri
(d) Jonhson

30. Malphigian Tubules are absent in

(a) Moths
(b) Butterfly
(c) Silk worm
(d) Aphids

31. Malphigian Tubules are 5 in number

(a) Housefly
(b) Stylopids
(c) Mosquitoes
(d) Bugs

32. The ringed appearance of the trachea is due to the spiral ridges called

(a) Peritrime
(b) Taenidia
(c) Both
(d) None of the above

33. Bursicon is a hormone that promotes

(a) Vitellogenesis
(b) Apolysis
(c) Tanning
(d) Moulting

34. Trachea is a invagination of

(a) Ectoderm
(b) Endoderm
(c) Mesoderm
(d) None of the above

35. Amino acid essential for scelorotisation is

(a) Bursicon
(b) Tyrosine
(c) Proline
(d) All of the above

36. Vitamins absent in insect body system

(a) A and B
(b) A and D
(c) C and K
(d) D and K

37. True salivary glands of hymenoptera

(a) Labial glands (b) Maxilary glands

(c) Lateral pharyngeal glands (d) Mandibular glands

38. Most common inorganic anion in insect blood

(a) Chloride (b) Sodium

(c) Potassium (d) Magnesium

39. Most common essential amino acid in insect

(a) Proline (b) Glycine

(c) Aspartic acid (d) Histidine

40. Adepokinetic hormone is secreted by

(a) Corpora Cardiaca (b) Corpora Allata

(c) Pro Thoracic Gland (d) Brain

41. Resting potential is also known as

(a) Polarised state (b) Depolarised state

(c) Excited state (d) Inert state

42. The dipteran which is luminiferous in nature due to malphigian tubule

(a) Megalagrian (b) Protodonat

(c) Bolitophila (d) Trigopolytus

43. In polymorphism of aphids first firmed generation is called

(a) Fundatrix (b) Brachypterous

(c) Macropterous (d) None of the above

44. Chiranomids respire through

(a) Blood gills (b) Tracheal gills

(c) Anal gills (d) Rectal gills

45. Ecdysone is secreted by

(a) Corpora Cardiaca (b) Corpora Allata

(c) Prothoracic Gland (d) Neuro Secretary Cells

46. The main function of filter chamber is

(a) To filter the food material

(b) To filter the air during respiration

(c) To filter the excess water directly to hind gut

(d) To filter the blood

47. Ecdysone is chemically

(a) Carbohydrate and fat
(b) Protein and carbohydrate
(c) Fat and protein
(d) Steroid

48. Cryptonephridial condition occurs in

(a) Diptera
(b) Siphonaptera
(c) Dermaptera
(d) Coleoptera

49. In winged insects, dilated trachea are known as

(a) Plastron
(b) Air sac
(c) Gills
(d) Hydrofuse hairs

50. Outer wall of the egg is known as

(a) Vitelline membrane
(b) Periplasm
(c) Chorion
(d) Blastoderm

ANSWER KEY

1	2	3	4	5	6	7	8	9	10	11	12	13	14	15
(b)	(a)	(b)	(b)	(a)	(a)	(d)	(a)	(c)	(a)	(d)	(d)	(d)	(c)	(c)
16	17	18	19	20	21	22	23	24	25	26	27	28	29	30
(b)	(a)	(b)	(b)	(b)	(d)	(b)	(a)	(c)	(b)	(c)	(b)	(c)	(a)	(d)
31	32	33	34	35	36	37	38	39	40	41	42	43	44	45
(c)	(b)	(c)	(a)	(b)	(d)	(a)	(a)	(c)	(c)	(a)	(c)	(a)	(a)	(c)
46	47	48	49	50										
(c)	(d)	(d)	(b)	(c)										

Chapter 26

Insect Growth Regulators and Pest Management, IPM Practices, Cultural, Mechanical, Biological, Chemical, Physical Methods, Semiochemicals, Ecology, Mode of Entry, Mode of Action, Carbamates, Ops, Organochlorins, Toxicology, Economic Decision Levels

1. **Application of superphosphate @ 80kg/ha to overcome ______ pest in rice crop**

   (a) Rice mealy bug (b) Rice root weevil
   (c) Stem borer (d) All of the above

2. **Destruction of weed host is practiced for the management of**

   (a) Citrus fruit sucking moth (b) Rice earhead bug
   (c) Both (a) and (b) (d) None of the above

3. **Madras Agricultural pest and disease act started in the year**

   (a) 1912 (b) 1914
   (c) 1919 (d) 1968

4. **Flooding the nursery is to control**

   (a) Swarming caterpillar (b) Army worm
   (c) Thrips (d) Both (a) and (b)

5. **Alternate wetting and drying of paddy field is method of pest control**

   (a) Physical (b) Chemical
   (c) Mechanical (d) Cultural

6. **Mixing crude oil in irrigation water is best method for management of**
   (a) Root grub (b) Cutworms
   (c) Root aphid (d) Ground beetle

7. **First quarantine act in the USA came into force in**
   (a) 1805 (b) 1905
   (c) 1705 (d) 2005

8. **The first state to enact agricultural pests and diseases act in 1919 is**
   (a) Andhra Pradesh (b) Tamil Nadu
   (c) Madya Pradesh (d) Uttar Pradesh

9. **The insecticide act was enacted in the year**
   (a) 1968 (b) 1969
   (c) 1972 (d) 2008

10. **Where are NCIPM and NCPHM located respectively?**
    (a) Bangalore and Delhi (b) Delhi and Hyderabad
    (c) Delhi and pune (d) Hyderabad and Delhi

11. **Pesticide management bill was passed in the year**
    (a) 2007 (b) 2008
    (c) 2006 (d) 2009

12. **Egg-laying by bruchids on pulses is prevented by**
    (a) Fumigation (b) Applying dust
    (c) Sun drying (d) Oil treatment

13. **Establishment Year of Directorate of Plant Protection, Quarantine and Storage**
    (a) 1945 (b) 1955
    (c) 1946 (d) 1958

14. **Destructive insect pests act enacted in the year**
    (a) 1914 (b) 1924
    (c) 1916 (d) 1968

15. **Dipel and Halt are the available forms of**
    (a) *Bacillus thuringiensis* (b) Fungi
    (c) Virus (d) Bacteria

16. Application of chemical at dusk time

(a) Stem borer
(b) Whitefly
(c) Cutworm
(d) White grubs

17. NBAIR is located at

(a) Chennai
(b) Hyderabad
(c) Bangalore
(d) New Delhi

18. NCIPM is located at

(a) Bhubaneshwar
(b) Cuttack
(c) Coimbatore
(d) New Delhi

19. Pesticide umbrella is a system of pest management is given by

(a) Pedigo
(b) Pradhan
(c) Kogan
(d) Smith

20. Increse in population of non targeted arthropod after pesticide application is

(a) Resistence
(b) Resurgence
(c) Secondary outbreak
(d) None of the above

21. Ferrolure produced by redpalm weevil is

(a) Sex pheromone
(b) Trail pheromone
(c) Aggregation pheromone
(d) Alarm pheromone

22. Pesticide bill (year)

(a) 1968
(b) 1971
(c) 2005
(d) 2008

23. Puddling is comman practice in rice for management of

(a) Caseworm
(b) BPH
(c) Mealybug
(d) Root aphid

24. Maintanence of optimum plant density in rice is management for

(a) BPH
(b) GLH
(c) WBPH
(d) None of the above

25. DDT is toxic to plant

(a) Cucurbits
(b) Cereals
(c) Tomato
(d) Both (a) and (c)

26. Particle size of Aerosol or Fog ranges between

(a) 0.001-50u (b) 0.0001-1u

(c) 1-50u (d) 0.01-25u

27. The sprayer is used for spraying tall plantation trees

(a) Bucket sprayer (b) Rocker sprayer

(c) Knapsack sprayer (d) Plunger sprayer

28. Quantity of insecticidal solution required for ULV sprays is

(a) <200L (b) <5L

(c) 200-400L (d) <400L

29. Pheromones are.in origin

(a) Endohormone (b) Exohormone

(c) Ectohormone (d) Both (b) and (c)

30. First sex pheromone in the world is

(a) Helilure (b) Bombykol

(c) Spodolure (d) Gyplure

31. Males of bark beetle secrete type of pheromone

(a) Sex (b) Aggregation

(c) Trial marking (d) Spacing

32. Citronollol is used as against mosquitoes and Japanese beetle

(a) Antifeedent (b) Repellent

(c) Antimetabolite (d) Attractant

33. The first synthetic chemical repellent used against insects

(a) Z.I.P (b) Creosote

(c) Bordeux mixture (d) Triazines

34. Arprocarb (Baygon) is used as repellent against

(a) Cotton boll worm (b) Cotton boll weevil

(c) Cotton stem and shoot weevil (d) All of the above

35. Triphenyltins belongs to group

(a) Triazines (b) Organotins

(c) Phenyl group (d) Carbamate

36. Paris green is a double salt of

(a) Cu acetate and Cu acetate
(b) Cu acetate and Cu arsenite
(c) Cu arsenite and Cu arsenite
(d) None of the above

37. Nicotine is a.poison

(a) Stomach
(b) Nerve
(c) Contact
(d) Respiratory

38. The epoxide of Aldrin is

(a) Endrin
(b) Dieldrin
(c) Epo-aldrin
(d) Delta-aldrin

39. Insecticidal property of endosulphon was discoverd by. in the year 1956

(a) Thomas ray
(b) William Kirby
(c) W.Finkenbrink
(d) Ziedler

40. First synthetic analogue of pyrethrum is

(a) Alpha-permethrin
(b) Cyfluthrin
(c) Allethrin
(d) Lambda cyahalothrin

41. _______ is single dose anticoagulant

(a) Coumatetralyl
(b) Bromodiolone
(c) Warfarin
(d) Barium chloride

42. The Buprofezin is an

(a) Anticoagulent
(b) IGR
(c) Antimetabolite
(d) Repellent

43. Dowfume MC-2 is a trade name of

(a) HCN
(b) Methylbromide
(c) Bromomethane
(d) Both (b) and (c)

44. LC50 is determined by

(a) Probit analysis
(b) Potters tower test
(c) Both (a) and (b)
(d) Davids test of LC50

45. Lesser the value of LD 50, what does it indicate

(a) chemical is less toxic
(b) Chemical is highly toxic
(c) Chemical is synergist
(d) All of the above

46. AICRP on pesticide residue started in the year

(a) 1894 (b) 1984

(c) 1784 (d) 1874

47. The first notice of resistance of red scale to lime sulphur in for the first time

(a) Quayle (b) Melander

(c) Garyson (d) Cochran

48. First agriculture pest showed resistance to DDT and HCH in India in 1963 is

(a) *S. litura* (b) DBM

(c) *H.armigera* (d) *G.birmanica*

49. The first insect showing resistance to DDT in India in 1952 is

(a) *Anapheles spp* (b) *Culex fatigans*

(c) *Aedes aegypti* (d) All of the above

50. In insecticidal formulations, EC means

(a) Effective concentration

(b) Equal concentration of ingredients

(c) Estimated chemical toxicity

(d) Emulsified concentrate

ANSWER KEY

1	2	3	4	5	6	7	8	9	10	11	12	13	14	15
(c)	(c)	(c)	(d)	(a)	(c)	(b)	(b)	(a)	(b)	(b)	(d)	(c)	(a)	(a)
16	17	18	19	20	21	22	23	24	25	26	27	28	29	30
(c)	(c)	(d)	(b)	(c)	(b)	(d)	(c)	(a)	(d)	(c)	(b)	(b)	(d)	(b)
31	32	33	34	35	36	37	38	39	40	41	42	43	44	45
(b)	(b)	(c)	(b)	(b)	(b)	(c)	(b)	(c)	(b)	(b)	(b)	(c)	(c)	(b)
46	47	48	49	50										
(b)	(b)	(d)	(a)	(d)										

Chapter 27

Insect Taxonomy, Mites, Endopterygotes and Exopterygotes, Families from each Orders and their Characters

1. **Until 2002, the most recent insect order was Grylloblattodea described in 1914. But in 2002, a new order was described with the name**
   (a) Mantophasmatodea
   (b) Phasmtomantodea
   (c) Gryllophasmotodea
   (d) Mantogryllodea

2. **An important object of the ICZN is to provide**
   (a) Uniqueness to the scientific names
   (b) Stability to the scientific names
   (c) Distinctiveness to the scientific names
   (d) All the three of the above

3. **Sub-social insects living in silken tunnels are**
   (a) Earwigs
   (b) Embiids
   (c) Book lice
   (d) Rock crawlers

4. **The larva construct pit fall traps in**
   (a) Owl flies
   (b) Aphid lions
   (c) Ant lions
   (d) Mantid flies

5. **In agaontid wasp males are**
   (a) Winged
   (b) Wingless
   (c) Winged or wingless
   (d) Winged but sometimes wingless

6. **In syrphids**
   (a) Larvae are predators
   (b) Adults are pollination
   (c) Both larvae and adults
   (d) Both (a) and (b)

7. **Caddis flies belong to the order**
   (a) Lepidoptera
   (b) Diptera
   (c) Thysanoptera
   (d) Trichoptera

8. **Scorpion flies are**
   (a) Plecoptera
   (b) Mecoptera
   (c) Coleoptera
   (d) Diptera

9. **Physapoda is another name of**
   (a) Isoptera
   (b) Coleoptera
   (c) Thysanoptera
   (d) Diptera

10. **Spurious veins are present in**
    (a) Robber flies
    (b) Hover flies
    (c) Both (a) and (b)
    (d) None of the above

11. **In collembolan the body segments divided into**
    (a) 9
    (b) 3
    (c) 6
    (d) 11

12. **Which of the following is an youngest order**
    (a) Grylloblattodea
    (b) Diptera
    (c) Zoraptera
    (d) Hemiptera

13. **How many pairs of legs are present in the class Aracnida**
    (a) 5 pairs
    (b) Two pairs
    (c) Four pairs
    (d) (d)Three pairs

14. **Leg bearing two segmented trochanter present in**
    (a) Odonata
    (b) Ichneumonidae
    (c) Both (a) and (b)
    (d) Strepsiptera

15. **Eye fly *Siphunculina funicola* belongs to the family**
    (a) Psychodidae
    (b) Chloropidae
    (c) Drosophilidae
    (d) Muscidae

16. Fringed wings present in

(a) Plume moth/pterophoridae (b) Thrips/Thysanoptera

(c) Both (a) and (b) (d) None of the above

17. Short and unsegmented type of cerci present in

(a) May fly (b) Grass hopper

(c) Cricket (d) Cockroach

18. The following order closely related to Lepidoptera is

(a) Strepsiptera (b) Diptera

(c) Trichoptera (d) Neuroptera

19. The following insect is dorsolaterally compressed

(a) Flea (b) Headlouse

(c) Dust lice (d) Biting lice

20. In sawflies _______ pairs of prolegs are present

(a) 4 pairs (b) 5 pairs

(c) 6-8 pairs (d) 8 pairs

21. Which insect is symbol of beauty, purity of soul

(a) Butterfly (b) Fly

(c) Honeybee (d) Beetle

22. Tail like prolongation of hind wings in papalionidae is

(a) M1vein (b) M2 vein

(c) M3 vein (d) All of the above

23. The insect order most closely related to Diptera is

(a) Orthoptera (b) Plecoptera

(c) Hymenoptera (d) Thysenoptera

24. Which order is not holometabolous

(a) Siphenaptera (b) Hemiptera

(c) Neuroptera (d) Diptera

25. Which order is exclusively herbivorous

(a) Trichoptera (b) Hemiptera

(c) Thysenoptera (d) Phasmatodea

26. Which order is exclusively endoparasitic

(a) Phthiraptera (b) Hymenoptera

(c) Strepsiptera (d) Diptera

27. Which order has both holo and Hemimetabolous insects

(a) Hymenoptera (b) Diptera

(c) Strepsiptera (d) Siphenaptera

28. Sucking mouthparts are not found in

(a) Fleas (b) Lice

(c) Ants (d) Bugs

29. All ametabolous insects are

(a) Predatory (b) Wingless

(c) Endognathous (d) Herbivorous

30. Immatures of Neuroptera can be classified as

(a) Scavengers (b) Parasites

(c) Predators (d) Herbivores

31. What do the orders Mantodea, Dermaptera and isopteran have in common

(a) Social life (b) Winglessness

(c) Well developed Cerci (d) chewing mouthparts

32. Which insect order is most closely related to Dermaptera

(a) Hymenoptera (b) Orthoptera

(c) Plecoptera (d) Thysanoptera

33. Chewing mouthparts never occur in

(a) Fleas (b) Beetles

(c) Bees (d) Earwigs

34. All neopterous insects are

(a) Predatory (b) Ectognathous

(c) Wingless (d) Hemimetabolous

35. Which character would not be found in class Onychophora

(a) One pair of antenna (b) 3 tagmata

(c) Jointed legs with claw (d) Segmented body

36. Which Arthropods have chelicerae

(a) Millipedes (b) Spiders

(c) Shrimp (d) Onychophora

37. Which statement is true about all Crustaceans

(a) Live on land (b) Have chewing mouthparts

(c) Six walking legs (d) Paleopterous

38. Tarsus single segmented in case of

(a) Clasping type (b) Fossorial type

(c) Scansorial type (d) Natatorial type

39. The taxonomy concerned with classification of species

(a) Beta taxonomy (b) Gama taxonomy

(c) Alpha taxonomy (d) None of the above

40. The taxonomy concerned with evolutionary relations and phylogeny

(a) Beta taxonomy (b) Gama taxonomy

(c) Alpha taxonomy (d) None of the above

41. The hierarchical concept of classification given by

(a) Carl von Linnaeus (b) Lamark

(c) Darwin (d) None of the above

42. The basic taxonomic category is

(a) Species (b) Sub species

(c) Genera (d) family

43. The lower most category for classifying an organism is

(a) Species (b) Sub species

(c) Genera (d) Family

44. Mandibular articulation in head is monocondylic

(a) Apterygota (b) Pterygota

(c) both (d) None of the above

45. The wing flexing mechanism is absent in

(a) Paleoptera orders (b) Panorpoid orders

(c) Hemipteroid orders (d) Orthopteroid orders

46. The insect group have the haustellate mouth parts and feed on liquid food

(a) Paleopteran order (b) Panorpoid complex

(c) Hemipteroid (d) Orthopteroid orders

47. The insect group have larval pupal stages in their life cycle

(a) Paleopteran order
(b) Panorpoid complex
(c) Hemipteroid
(d) Orthopteroid orders

48. Apterygota insect continues its moulting even after attaining sexual maturity

(a) Ephemeroptera
(b) Thysanura
(c) Hymenoptera
(d) Coleoptera

49. The sub imago stage present in

(a) Dragon fly
(b) Damsel fly
(c) Mayfly
(d) All of the above

50. Pterigotan insect order continues its moulting even after becoming adult

(a) Ephemeroptera
(b) Odonata
(c) Hymenoptera
(d) Coleoptera

ANSWER KEY

1	2	3	4	5	6	7	8	9	10	11	12	13	14	15
(a)	(d)	(b)	(c)	(b)	(d)	(d)	(b)	(c)	(b)	(c)	(d)	(c)	(c)	(b)
16	17	18	19	20	21	22	23	24	25	26	27	28	29	30
(b)	(b)	(c)	(a)	(d)	(a)	(c)	(c)	(b)	(c)	(c)	(c)	(c)	(b)	(c)
31	32	33	34	35	36	37	38	39	40	41	42	43	44	45
(d)	(b)	(a)	(b)	(d)	(b)	(b)	(c)	(a)	(b)	(a)	(a)	(b)	(a)	(a)
46	47	48	49	50										
(c)	(b)	(b)	(c)	(a)										

Chapter 28

Beneficial Insects (Lac, Honeybees, Silkworm), different Agricultural and Horticultural Pests and their Management

1. **Per cent of sucrose and fructose present in honey is**
   (a) 41 per cent and 2 per cent
   (b) 2 per cent and 41 per cent
   (c) 35 per cent and 12 per cent
   (d) 12 per cent and 35 per cent

2. **K diseases in honeybees is caused by**
   (a) TSBV
   (b) American foul brood
   (c) Mites
   (d) European foul brood

3. **Silk glands are modified**
   (a) Labial glands
   (b) Salivary glands
   (c) Maxillary glands
   (d) Labral glands

4. **All India bee keepers association is located at**
   (a) Pune
   (b) Bangalooru
   (c) Hisar
   (d) Delhi

5. **Scientific name of Oak Tasar silkworm is**
   (a) *Antherea assamensis*
   (b) *Antherea mylitta*
   (c) *Antherea proylei*
   (d) *Antherea javanensis*

6. **The best quality of silk is obtained from**
   (a) Multivoltines
   (b) Bivoltines
   (c) Univoltines
   (d) Independent of voltinism

7. silk is called as Ahimsa silk

(a) Mulberry (b) Tasar

(c) Eri (d) Muga

8. Shellac is manufactured by. process

(a) Solute extraction (b) Solvent extraction

(c) Bayers process (d) All of the above

9. The length of proboscis of worker honey bee is

(a) 3-5 mm (b) 4mm

(c) 6mm (d) 6-8mm

10. The two Strains of Rangeeni lac are

(a) Aghani and Jethwi (b) Katki and Baisaki

(c) Aghani and Baisaki (d) Katki and Jethwi

11. Highest lac producing state is

(a) Tamil nadu (b) Kerala

(c) Jharkhand (d) Odisha

12. Which of the following is a chemosterilent?

(a) Amithopterin (b) 5-Bromouracil

(c) Methohexate (d) (d)All of the above

13. Honey bee worker is

(a) Functional male (b) Functional female

(c) Imperfectly developed female (d) Both (a) and (b)

14. Silk fibre is made up of inner fibrion and outer sericin at the ratio of

(a) 1 : 3 (b) 2 : 2

(c) 3 : 1 (d) 1 : 1

15. Silk produced by ______ gland

(a) Labial (b) Mandibular

(c) Salivary (d) All of the above

16. The newly hatched silkworms are known as

(a) Ants (b) Kegs

(c) Kengos (d) All of the above

17. Royal jelly is secreted by

(a) Maxillary glands
(b) Labial glands
(c) Mandibular glands
(d) Hypopharyngeal glands

18. In honey bees the meta thorax is fused with the I abdominal segment called

(a) Gaster
(b) Propodium
(c) Petiole
(d) Sting

19. Poison glands are associated with

(a) Alkaline glands
(b) Stinging organ
(c) Cephalic glands
(d) Both (a) and (b)

20. Muga silk in India is obtained from

(a) Uttar Pradesh
(b) Karnataka
(c) Orissa
(d) Assam

21. Silk gland in silkworm is a modification of

(a) Salivary gland
(b) Mandibular gland
(c) Maxillary gland
(d) Pharyngeal gland

22. The most prevalent domesticated silkworm in India is

(a) Mulberry silkworm
(b) Eri silkworm
(c) Muga silkworm
(d) Tassar silkworm

23. The safest insecticide for honeybee is

(a) Malathion
(b) Chlorpyriphos
(c) Endosulfan
(d) DDVP

24. Lac production more in

(a) Bihar
(b) Karnataka
(c) Madhy Pradesh
(d) Rajasthan

25. The major protein found in silk is

(a) Fibroin
(b) Elastin
(c) Sericin
(d) Keratin

26. Damaging stage of Epilachna beetle is

(a) Grub only
(b) Adult only
(c) Nymph and adult
(d) Grub and adult

27. Which instar larva of citrus butterfly looks like birds dropping

(a) Late instar (b) Early instar
(c) Pupal stage (d) None of the above

28. Little leaf of brinjal is caused by

(a) Mycoplasma (b) Phytoplasma
(c) Bacteria (d) Virus

29. Cecidogen is secreted by instar larvae of *Orseolia oryzae*

(a) 1st instar (b) 2nd instar
(c) 3rd instar (d) Both (a) and (b)

30. Bad boll opening in cotton due to

(a) Aphids (b) Whitefly
(c) Pink bollworm (d) American bollworm

31. Cylas formicarius belongs to family

(a) Curculionidae (b) Apidae
(c) Apionidae (d) Anodibae

32. Golden yellow mosaic disease of bhendi transmitted by

(a) Aphids (b) Thrips
(c) Leafhopprs (d) Whitefly

33. Heavy shedding of early formed flowers buds of cotton due to

(a) American bollworm (b) Pinkbolworm
(c) Spotted bollworm (d) Cottom aphid

34. White grub belongs to family

(a) Scarabidae (b) Carabidae
(c) Melolonthidae (d) Cerambicidae

35. "Silver shoot" or onion shoot is modified

(a) Main stem (b) Petiole
(c) Leaf margin (d) Leaf sheath

36. The eggs laid in the masses near the tip of tender leaf blade

(a) *Scirpophaga incertulas* (b) *Scirpophaga excerptalis*
(c) Top shoot borer (d) Rice case worm

37. Kerala wilt in coconut is transmitted by

(a) Thrips (b) Scales

(c) Mealy bug (d) Tingid

38. Onion fly *Hylemna antiqui* belongs to family

(a) Cecidomydae (b) Tachinidae

(c) Muscidae (d) None of the above

39. Which of the following is most effective chemical against sugarcane woolly aphid

(a) Chlorpyriphos (b) Thimethaxoam

(c) Monocrotophos (d) None of the above

40. DD – 136 is effective against

(a) Lepidoptera (b) Coleoptera

(c) Hemiptera (d) All of the above

41. Fruit flies pupate in

(a) On stem (b) Inside fruits

(c) Soil (d) Leaf litter

42. Citrus tristiza is transmitted by

(a) *Orocius albicinctus* (b) *Cestius phyctis*

(c) *Toxoptera citricida* (d) *Brachycadus helichrysi*

43. Turnip yellow mosaic disease transmitted by

(a) Aphididae (b) Cicadellidae

(c) Forficulidae (d) Thripidae

44. Rice yellow dwarf virus transmitted by

(a) Green Plant Hopper (b) Brown Plant Hopper

(c) White backed BPH (d) Green leaf bug

45. Entry of exotic pests can be checked through

(a) Chemical control (b) Physical barriers

(c) Repellent usages (d) Quarantine measures

46. Ber fruits are attacked by

(a) *Batocera* spp. (b) *Dichocrosis* spp.

(c) *Carpomyia vesuviana* (d) *Achoea janata*

47. *Bagrada hilaris* is a serious pest of

(a) Mustard
(b) Rice
(c) Red gram
(d) Sugarcane

48. Young leaves of citrus plants show serpentine galleries due to

(a) *Diaphorina citri*
(b) *Phyllocnistis citrella*
(c) *Dialeurodes citri*
(d) *Aleurocanthus woglumi*

49. The vector of tomato spotted wilt virus causing bud necrosis in groundnut

(a) Pod sucking bug
(b) Leaf miner
(c) Aphids
(d) Thrips

50. A serious pest of sorghum earheads is

(a) *Chilo partellus*
(b) Wingless grasshopper
(c) *Pyrilla*
(d) Midge

ANSWER KEY

1	2	3	4	5	6	7	8	9	10	11	12	13	14	15
(b)	(c)	(a)	(a)	(c)	(c)	(c)	(d)	(a)	(b)	(c)	(d)	(c)	(c)	(c)
16	17	18	19	20	21	22	23	24	25	26	27	28	29	30
(d)	(c)	(b)	(d)	(d)	(a)	(a)	(c)	(a)	(a)	(d)	(b)	(a)	(d)	(c)
31	32	33	34	35	36	37	38	39	40	41	42	43	44	45
(c)	(d)	(c)	(a)	(d)	(a)	(c)	(c)	(c)	(d)	(c)	(c)	(c)	(a)	(d)
46	47	48	49	50										
(c)	(a)	(b)	(d)	(d)										

SOCIAL SCIENCES

Chapter 29

Introduction of Extension, Extension Education, Types of Education, Informal, Formal, Non-formal, Classification of Communication, Individual, Group, Mass, Developmental Programmes of Pre-independence and Post-independence Era, Watershed Developmental Programmes

1. **Extension is a _______?**
   (a) Two-way flow of message
   (b) 1 way flow of message
   (c) Both (a) and (b)
   (d) Multi-way flow of message

2. **Which is known as package programme?**
   (a) IRDP
   (b) CADP
   (c) IADP
   (d) DPAP

3. **The role of extension education in India is performed by.?**
   (a) State Agricultural University
   (b) Central Agricultural University
   (c) Both (a) and (b)
   (d) Krishi Vigyan Kendra

4. **Motivation is a process of _______?**
   (a) Enhancing the skill of individuals by spring them through the examples of various success stories
   (b) Initiating a conscious and purposeful action

(c) To motivate laggards to get something extraordinary
(d) All of the above

5. **By nature, extension teaching is _______?**
(a) Horizontal
(b) Vertical
(c) Vertical – horizontal
(d) None of the above

6. **Over adoption may occur due to _______?**
(a) Sufficient knowledge
(b) More knowledge
(c) Insufficient knowledge
(d) No knowledge

7. **What is the early adopter's percentage?**
(a) 2.5
(b) 13.5
(c) 16
(d) 10

8. **The sensory situation is more in _______?**
(a) Interpersonal communication
(b) Intrapersonal communication
(c) Both (a) and (b)
(d) None of the above

9. **Who conceptualized management as the process of designing and maintaining an environment in which individuals working together in groups.**
(a) Daniel Bonor
(b) Hamilton
(c) Leagans
(d) Knootz and Weihrich

10. **What is the meaning of Esprit de corps?**
(a) Individual sprit
(b) Work sprit
(c) Outcome sprit
(d) Team sprit

11. **The major activities of agricultural extension division are _______?**
(a) Technology assessment
(b) Demonstration
(c) Capacity development
(d) All of the above

12. **Who is the deputy director general of agricultural extension of India?**
(a) Dr. A. K. Singh
(b) Dr. R. K. Varshney
(c) Shri N. S. Tomar
(d) Dr. Trilochan Mohapatra

13. **ATARIS play the role of _______?**
(a) Extension of research works
(b) Coordination and monitoring with headquarters

(c) Collaboration with extension personnel and farmers
(d) All of the above

14. Total number of kvks is more under?
(a) Non-government Organization
(b) State agricultural university
(c) ICAR institutes
(d) Central agricultural institutes

15. Sevagram project was started by ______ In the ______?
(a) Mahatma Gandhi, 1949
(b) Rabindranath Tagore, 1942
(c) Mahatma Gandhi, 1933
(d) Rabindranath Tagore, 1932

16. Which pair is right ______?
(a) Nilokheri project – Albert Mayer
(b) Etawah project – Spencer hatch
(c) Shantiniketan – R. Tagore in collaboration with Leonard K
(d) Gurgaon project – Spencer hatch in collaboration with Mahatma Gandhi

17. The origin place of the modern extension is ______?
(a) Moscow, Russia
(b) Dublin, Ireland
(c) Kassel, Germany
(d) Brussels, Belgium

18. The ideal size of a self-help group is ______?
(a) 5 – 10
(b) 10 – 15
(c) 10 – 25
(d) 15 – 30

19. Which is the first step of extension educational process?
(a) Collection of facts
(b) Analysis of the situations
(c) Both (a) and (b)
(d) Deciding on realistic objectives

20. Which is a principle of extension education?
(a) Principles of grass root level organisation
(b) Cooperation and people's participation
(c) Indigenous knowledge
(d) All of the above

21. TRYSEM was launched in ______?

(a) 1956 (b) 1967

(c) 1971 (d) 1979

22. Father of extension in India

(a) A.K Singh (b) K.N Singh

(c) Hamilton (d) Trilochan Mohapatra

23. Model village was given by

(a) Spencer hatch (b) F.L brayne

(c) A.T Moscher (d) Daniel Hamilton

24. Gurgaon experiment was given by

(a) Spencer hatch (b) F.L Brayne

(c) A.T Moscher (d) Mahatma Gandhi

25. Etawah pilot project was started by

(a) Albert Mayer (b) Spencer hatch

(c) F.L Brayne (d) A.K Singh

26. Community Development Project was started in

(a) 1952 (b) 1953

(c) 1979 (d) 1978

27. National Extension Service was started in

(a) 1952 (b) 1953

(c) 1978 (d) 1979

28. Panchayati raj was started in

(a) 1953 (b) 1952

(c) 1958 (d) 1978

29. IRDP, TRYSEM, DWACRA, SITRA merged into

(a) SSGY (b) PMRY

(c) JGSY (d) SGSY

30. Lab to land program started in

(a) 1958 (b) 1959

(c) 1978 (d) 1979

31. First five-year plan started in

(a) 1951 (b) 1946

(c) 1956 (d) 1961

32. Agriculture Research Services were established by

(a) Ms swaminathan (b) Hamilton

(c) Spencer hatch (d) F.L Brayne

33. Soil health card was launched for the first time in

(a) Rajasthan (b) Punjab

(c) Haryana (d) Telengana

34. First KVK was recommended by

(a) Balwant Rai Mehta (b) Mohan Singh Mehta

(c) Ashok Mehta (d) None of the above

35. First K.V.K was established at

(a) Pondicherry (b) Goa

(c) Uttar Pradesh (d) Madhya Pradesh

36. T and V program was adopted first in

(a) Rajasthan (b) Uttar Pradesh

(c) Madhya Pradesh (d) None of the above

37. First bharatiya mahila Bank was launched in

(a) New Delhi (b) Mumbai

(c) Chennai (d) Kolkata

38. First state to adopt panchayati raj

(a) Rajasthan (b) Uttar Pradesh

(c) Madhya Pradesh (d) Maharashtra

39. Firka development schemes was given by

(a) T. Prakasham (b) S.K Dey

(c) Albert Mayer (d) Spencer hatch

40. Mazdoor Manzil was given by

(a) T. Prakasham (b) S.K Dey

(c) Albert Mayer (d) Spencer hatch

41. Marthandan project was given by

(a) T. Prakasham (b) S.K Dey

(c) Albert Mayer (d) Spencer hatch

42. The state which was electrified first is

(a) Haryana (b) Punjab

(c) Uttar Pradesh (d) Madhya Pradesh

43. The right approach for agriculture extension is

(a) Top to bottom approach
(b) Bottom to top approach
(c) Horizontal approach
(d) None of the above

44. ISOPOM was established in the year

(a) 2002
(b) 2004
(c) 2006
(d) 2008

45. Grow more food campaign was given in

(a) 1945
(b) 1948
(c) 1952
(d) 1956

46. Which country has developed an artificial intelligence Nanny (AI Nanny), to monitor embryos?

(a) Japan
(b) China
(c) USA
(d) Germany

47. Lake Albert' is located as a border between which two countries?

(a) Uganda-Democratic Republic of Congo
(b) Algeria-Libya
(c) South Africa- Tanzania
(d) Uganda- Ethiopia

48. Which Union Ministry implements 'Digital India Land Record Modernisation Programme (DILRMP)?

(a) Ministry of Electronics and IT
(b) Ministry of Rural Development
(c) Ministry of Home Affairs
(d) Ministry of Statistics and Programme Implementation

49. Which company has started India's first project of mixing Hydrogen into Natural Gas System?

(a) ONGC
(b) GAIL
(c) NTPC
(d) Power Grid Corporation

50. Which country's researchers have found out that Climate change is driving plants to flower earlier?

(a) India
(b) USA
(c) UK
(d) Australia

ANSWER KEY

1	2	3	4	5	6	7	8	9	10	11	12	13	14	15
(a)	(c)	(a)	(b)	(a)	(c)	(b)	(b)	(d)	(d)	(d)	(a)	(b)	(b)	(c)
16	17	18	19	20	21	22	23	24	25	26	27	28	29	30
(c)	(d)	(d)	(d)	(d)	(d)	(b)	(d)	(b)	(a)	(a)	(b)	(c)	(a)	(d)
31	32	33	34	35	36	37	38	39	40	41	42	43	44	45
(a)	(a)	(a)	(b)	(a)	(a)	(b)	(a)	(a)	(b)	(d)	(a)	(b)	(b)	(b)
46	47	48	49	50										
(b)	(a)	(b)	(b)	(c)										

Chapter 30

TRYSEM, IRDP, DWCRA, JRY, SGSY, ATMA, Projects and Associated Persons, Audio Visual Aids, Factors Influencing Learning, Teaching Steps, Communication Models, Categories of Adaptors

1. **The percentage of innovators in new innovation is**
   (a) 2.5 per cent (b) 13.5 per cent
   (c) 34 per cent (d) 16 per cent

2. **The percentage of early adaptor is how much per cent of total population**
   (a) 2.5 per cent (b) 13.5 per cent
   (c) 34 per cent (d) 16 per cent

3. **The percentage of early majority and late majority is**
   (a) 2.5 per cent (b) 13.5 per cent
   (c) 34 per cent (d) 16 per cent

4. **The percentage of laggards in the new adopters is**
   (a) 2.5 per cent (b) 13.5 per cent
   (c) 34 per cent (d) 16 per cent

5. **Extension is derived from**
   (a) Latin word (b) Greek word
   (c) French word (d) None of the above

6. **MANAGE is established in the year**
   (a) 1980 (b) 1986
   (c) 1990 (d) 1996

7. **Father of extension in India**
   (a) S K dey (b) K.N singh
   (c) P. Deshmukh (d) None of the above

8. **Concept of demonstration is started by**
   (a) Dr Seaman A Knapp (b) Paul Leagans
   (c) T moscher (d) None of the above

9. **Which of the following is known as package programme**
   (a) IADP (b) SGSY
   (c) IAAP (d) CACP

10. **Extension is a**
    (a) One way flow of message (b) Two-way flow of message
    (c) Three-way flow of message (d) None of the above

11. **The nationalization of bank is done in**
    (a) 1969 (b) 1979
    (c) 1989 (d) 1999

12. **Lab to land programme is stared in**
    (a) 1969 (b) 1979
    (c) 1989 (d) 1999

13. **SGSY is started in the year**
    (a) 1969 (b) 1979
    (c) 1989 (d) 1999

14. **SGSY is representative of**
    (a) IRDP (b) TRYSEM
    (c) DWACRA, SITRA (d) All of the above

15. **Integrated rural development programme is started in**
    (a) 1968 (b) 1978
    (c) 1988 (d) 1998

16. **Training and Visit system is started by**
    (a) Daniel Benor (b) S A Knapp
    (c) S K DEY (d) Hamilton

17. T and V system started in

(a) 1964 (b) 1974
(c) 1984 (d) 1994

18. Green revolution is started in the year

(a) 1966 (b) 1960
(c) 1980 (d) 1990

19. High yielding variety programme is started in

(a) 1966 (b) 1980
(c) 1955 (d) 1970

20. IAAP (Intensive agriculture area programme) is started in

(a) 1954 (b) 1964
(c) 1974 (d) 1984

21. Intensive cattle development programme is started in

(a) 1963-64 (b) 1973-1974
(c) 1983-84 (d) None of the above

22. IADP is started in

(a) 1960 (b) 1950
(c) 1970 (d) 1980

23. Model village was given by

(a) T prakashan (b) P deshmukh
(c) Daniel benor (d) Hamilton

24. Father of extension

(a) Leagans (b) K N singh
(c) Hamilton (d) None of the above

25. National demonstratioon scheme is started by

(a) S K dey (b) T prakashan
(c) Kalwar and subramanian (d) R N gupta

26. Young farmer association is started by

(a) S K dey (b) P Deshmukh
(c) T prakashan (d) Kalwar and subramanian

27. Indian village scheme is given by

(a) S K Dey (b) T moscher and B N gupta
(c) T prakashan (d) Kalwar and subramanian

28. Firka movement was started by

(a) S K Dey (b) T Moscher

(c) B N Gupta (d) T Prakashan

29. Nilakhori experiment is done by

(a) S K Dey (b) T Prakashan

(c) T Moscher (d) P Deshmukh

30. Majdoor Manjil is started by

(a) S K Dey (b) T Prakashan

(c) T Moscher (d) P Deshmukh

31. National extension service is started in

(a) 1952 (b) 1953

(c) 1962 (d) 1963

32. Community development is started in

(a) 1952 (b) 1953

(c) 1962 (d) 1963

33. Etawah pilot project was started in which state

(a) Uttar Pradesh (b) Madhya Pradesh

(c) Andhra Pradesh (d) ODISHA

34. Etawah pilot project was started by

(a) Albert Mayer (b) Mahatma Gandhi

(c) Spencer Hatch (d) None of the above

35. Sevagram is started in

(a) Wardha (b) Varanasi

(c) Dandi (d) Ahmedabad

36. Sevagram project was started by

(a) M D Tagore (b) Mahatma Gandhi

(c) Specser Hatch (d) Deshmukh

37. Marthandam project was started by

(a) M D tagore (b) Mahatma Gandhi

(c) Spencer Hatch (d) Daniel Hamilton

38. Sriniketan project is started by

(a) Ravindra Nath tagore (b) Brayne

(c) Hamilton (d) Mahatma Gandhi

39. Sriniketan project was started in

(a) 1919
(b) 1921
(c) 1930
(d) 1931

40. Gurgaon project was started in the year

(a) 1920
(b) 1921
(c) 1930
(d) 1931

41. Gurgaon project is associated with

(a) F. L brayne
(b) R N tagore
(c) Hamilton
(d) Spencer hatch

42. Shantiniketan was started by

(a) F L Brayne
(b) M D Tagore
(c) R N Tagore
(d) Mahatma Gandhi

43. Group contacts methods are

(a) Symposium
(b) Discussion
(c) Seminar
(d) All of the above

44. Mass contact methods are

(a) Bulletin
(b) News paper
(c) Exhiobition
(d) All of the above

45. Individual contact methods are

(a) Farm and home visit
(b) Personal contact
(c) Telephone call
(d) All of the above

46. Which Union Ministry implements the Mahatma Gandhi National Rural Employment Guarantee Scheme (MGNREGS)?

(a) Ministry of Labour and Employment
(b) Ministry of Rural Development
(c) Ministry of Social Justice and Empowerment
(d) Ministry of Law and Justice

47. Which institution released a report on Climate Change that is dubbed as "atlas of human suffering"?

(a) World Economic Forum
(b) IPCC
(c) NITI Aayog
(d) WWF

48. Which state/UT hosted India's First Ice Wall Climbing Competition?

(a) Himachal Pradesh
(b) Ladakh
(c) Uttarakhand
(d) Sikkim

49. DIKSHA, SWAYAM and NISHTHA are initiatives of which Union Ministry?

(a) Ministry of MSME
(b) Ministry of Education
(c) Ministry of Skill Development
(d) Ministry of Environment, Forest and Climate Change

50. "Ungalil Oruvan" (One Among You) is the autobiography of the Chief Minister of which state?

(a) Kerala
(b) Tamil Nadu
(c) Odisha
(d) Telangana

ANSWER KEY

1	2	3	4	5	6	7	8	9	10	11	12	13	14	15
(a)	(b)	(c)	(d)	(a)	(b)	(b)	(a)	(a)	(b)	(a)	(b)	(d)	(d)	(b)
16	17	18	19	20	21	22	23	24	25	26	27	28	29	30
(a)	(b)	(a)	(a)	(b)	(a)	(a)	(d)	(a)	(c)	(b)	(b)	(d)	(a)	(a)
31	32	33	34	35	36	37	38	39	40	41	42	43	44	45
(b)	(a)	(a)	(a)	(a)	(b)	(d)	(a)	(b)	(a)	(a)	(b)	(d)	(d)	(d)
46	47	48	49	50										
(b)	(b)	(b)	(b)	(b)										

Chapter 31

Theories of Economics, Types of Farmers, Finance According to Period, Farm Efficiency Measures, Production Economics, Farm Management, Fixed Cost, Variable Cost, AGMARK, NABARD, Important Organizations, GST, CGST, SGST, IGST

1. **ECONOMICS is derived from which word?**
   (a) Greek (b) Latin
   (c) French (d) None of the above
2. **Theory of Profit is given by**
   (a) Walker (b) Hicks and hansen
   (c) Malthus (d) Lerner
3. **Morden theory of intrest is given by**
   (a) A walker (b) Hicks and Hansen
   (c) Malthus (d) Lerner
4. **Population theory is given by**
   (a) A. Walker (b) Hicks and Hansen
   (c) Malthus (d) A.P Lerner
5. **Theory of inflation is is given by**
   (a) Walker (b) Hicks and Hansen
   (c) Malthus (d) Lerner

6. **Theory of absolute advantage is given by**
   (a) Hicks and Hansen (b) Malthus
   (c) Lerner (d) Adam smith
7. **Marginal farmers have land holdings**
   (a) less than 1 ha (b) 1-2 ha
   (c) 2-4 ha (d) 4-10 ha
8. **Small farmers have land holding**
   (a) less than 1 ha (b) 1-2 ha
   (c) 2-4 ha (d) 4-10 ha
9. **Large farmers have land holding**
   (a) 1-2 ha (b) 2-4 ha
   (c) 4-10 ha (d) more than 10 ha
10. **Medium farmers have land holding**
   (a) 1-2 ha (b) 2-4 ha
   (c) 4-10 ha (d) 10 ha or more
11. **Crop loan is also known as**
   (a) Short term loan (b) Medium term loan
   (c) Long term loan (d) None of the above
12. **Time period for short term loan**
   (a) 1-1.5 years (b) 1.5-5 years
   (c) 5-30 years (d) None of the above
13. **Time period for medium term loan is**
   (a) 1-1.5 years (b) 1.5-5 years
   (c) 5-30 years (d) None of the above
14. **Time period for long term loan is**
   (a) 1-1.5 years (b) 1.5-5 years
   (c) 5-30 years (d) None of the above
15. **Total assets/Total liabilities is known as**
   (a) Net capital ratio (b) Current ratio
   (c) Debt equity ratio (d) None of the above
16. **Current assets/Current liabilities is known as**
   (a) Net capital ratio (b) Current ratio
   (c) Debt equity ratio (d) None of the above

17. **Differed liabilities/Net worth is known as**
(a) Net cxapital ratio (b) Current ratio
(c) Debt equity ratio (d) Production efficiency

18. **Total operation cost/total profit is known as**
(a) Production efficiency (b) Operating cost ratio
(c) Rate of capital turn over (d) None of the above

19. **Farm management comes under**
(a) Macro approach (b) Micro approach
(c) Inter farm approach (d) None of the above

20. **Production economics comes under**
(a) Macro approach (b) Inter farm approach
(c) Both (a) and (b) (d) Intra farm approach

21. **Fixed cost is also known as**
(a) Sunk cost (b) Variable cost
(c) Vary with level of output (d) None of the above

22. **GST collected by central government on an intrastate sale is**
(a) CGST (b) IGST
(c) SGST (d) None of the above

23. **GST collected by the state government on an intra state sale is**
(a) CGST (b) IGST
(c) SGST (d) None of the above

24. **GST collected by central government for inter state sale is**
(a) CGST (b) SGST
(c) IGST (d) None of the above

25. **GST came into effect from**
(a) 2015 (b) 2016
(c) 2017 (d) 2018

26. **GST is**
(a) Direct tax (b) Indirect tax
(c) Normal tax (d) None of the above

27. **NIAM is situated at**
(a) Jaipur (b) Raipur
(c) Hyderabad (d) Chennai

28. NIAM was set up on

(a) 1978 (b) 1988

(c) 1998 (d) 2008

29. WTO is situated at

(a) Geneva (b) California

(c) Berlin (d) Rome

30. WTO was set up on

(a) 1990 (b) 1995

(c) 2000 (d) 2005

31. Earlier name of WTO was

(a) TAGG (b) GATT

(c) WMO (d) IFM

32. WTO was founded during

(a) Uganda round (b) Uruguay round

(c) Paraguay round (d) None of the above

33. FCI was set up on

(a) 1975 (b) 1985

(c) 1965 (d) 1955

34. The head quarter of NABARD is

(a) Mumbai (b) Delhi

(c) Chennai (d) Bhubaneshwar

35. NABARD was established on

(a) 1962 (b) 1972

(c) 1982 (d) 1992

36. Present chairman of FCI is

(a) Sanjiv kumar (b) Deepak panwar

(c) R.S bhanwala (d) None of the above

37. Present chairman of NABARD is

(a) Dr. G R Chintala (b) Deepak panwar

(c) HB banwala (d) None of the above

38. The organization which gives indirect finance to rural community

(a) MANAGE (b) NABARD

(c) NIAM (d) IIM

39. Organization which promote a universal standard for facilitating international exchange of goods and services

(a) ISO
(b) BIS
(c) ISI
(d) Both (b) and (c)

40. International organization for standardization was set up on

(a) 1947
(b) 1957
(c) 1949
(d) 1959

41. Which mark provides third party guarantee to the buyer that goods are in accordance with the province of Indian standards

(a) BIS
(b) ISO
(c) AGMARK
(d) None of the above

42. The earlier name of BIS is

(a) ISI
(b) AGMARK
(c) ECOMARK
(d) ISO

43. Which mark is given ensure that product is environmentally friendly

(a) ECOMARK
(b) AGMARK
(c) BIS
(d) ISI

44. ECOMARK is starteds in

(a) 1991
(b) 1937
(c) 1987
(d) 1947

45. Which mark indicates the purity and quality of agriculture products

(a) AGMARK
(b) ECOMARK
(c) BIS
(d) ISI

46. AGMARK is set up on

(a) 1937
(b) 1991
(c) 1947
(d) 1958

47. Indian standard institution was established on

(a) 1987
(b) 1974
(c) 1947
(d) 1988

48. Which is the headquarters of the International Atomic Energy Agency (IAEA)?

(a) Geneva
(b) Vienna
(c) New York
(d) Moscow

49. Which city is the host of Mobile World Congress (MWC) 2022?

(a) Paris
(b) Barcelona
(c) Beijing
(d) Tokyo

50. 'National Gender Index' and 'State Energy and Climate Index' are the upcoming initiatives of which institution?

(a) UNDP
(b) NITI Aayog
(c) UNICEF
(d) Economist Intelligence Unit

ANSWER KEY

1	2	3	4	5	6	7	8	9	10	11	12	13	14	15
(a)	(a)	(b)	(c)	(d)	(d)	(a)	(b)	(d)	(c)	(a)	(a)	(b)	(c)	(a)
16	17	18	19	20	21	22	23	24	25	26	27	28	29	30
(b)	(c)	(b)	(b)	(c)	(a)	(a)	(c)	(c)	(c)	(b)	(a)	(b)	(a)	(b)
31	32	33	34	35	36	37	38	39	40	41	42	43	44	45
(b)	(b)	(c)	(a)	(c)	(a)	(a)	(b)	(a)	(a)	(a)	(a)	(a)	(a)	(a)
46	47	48	49	50										
(a)	(c)	(b)	(b)	(b)										

PLANT PHYSIOLOGY

Chapter 32

Plant Physiology, Diffusion, Osmosis, Plasmolysis, DPD, Solutions, Imbibitions, Water Relations, Mechanism of Ascent of Saps, Transpiration, Stomata, Guttation, Photosynthesis, Pigments of Photosynthesis, Light Reaction, Emerson Effect, Dark Reaction, Calvin Cycle, Law of Minimum

1. **Hormone primarily connected with cell division is**
   (a) IAA
   (b) NAA
   (c) Cytokinin/zeatin
   (d) Gibberellic acid

2. **Gibberellins promote**
   (a) Seed germination
   (b) Seed dormancy
   (c) Leaf fall
   (d) Root elongation

3. **The law of limiting factors was proposed by**
   (a) Leibig
   (b) Hatch and Slack
   (c) Blackman
   (d) Arnon

4. **Photorespiration is favoured by**
   (a) High temperature and low O_2
   (b) High humidity and temperature
   (c) High O_2 and low CO_2
   (d) High CO_2 and low O_2

5. **When water enters in roots due to diffusion, is termed as**
   (a) Osmosis
   (b) Passive absorption
   (c) Endocytosis
   (d) Active absorption

6. **Which of the following element plays an important role in biological nitrogen fixation?**
   (a) Copper
   (b) Molybdenum
   (c) Zinc
   (d) Manganese

7. **The closure of lid of pitcher in pitcher plant, is due to**
   (a) Paratonic movement
   (b) Autonomous movement
   (c) Turgor movement
   (d) Tropic movement

8. **The movement of water, from one cell of cortex to adjacent one in roots, is due to**
   (a) Accumulation of inorganic salts in the cells
   (b) Accumulation of organic compounds in the cells
   (c) Water potential gradient
   (d) Chemical potential gradient

9. **If the growing plant is decapitated, then**
   (a) Its growth stops
   (b) Leaves become yellow and fall down
   (c) Axillary buds are inactivated
   (d) Axillary buds are activated

10. **Removal of apical bud results in**
   (a) Formation of new apical bud
   (b) Elongation of main stem
   (c) Death of plant
   (d) Formation of lateral branching

11. **Translocation of carbohydrate nutrients usually occurs in the form of**
   (a) Glucose
   (b) Maltose
   (c) Starch
   (d) Sucrose

12. **Stomata open and close due to**
   (a) Circadian rhythm
   (b) Genetic clock
   (c) Pressure of gases inside the leaves
   (d) Turgor pressure of guard cells

13. Phytochrome is involved in

(a) Phototropism
(b) Photorespiration
(c) Photoperiodism
(d) Geotropism

14. Minerals absorbed by root move to the leaf through

(a) Xylem
(b) Phloem
(c) Sieve tubes
(d) None of the above

15. The size of chlorophyll molecule is

(a) Head 15 x 15 A, tail 25 A
(b) Head 20 x 20 A, tail 25 A
(c) Head 15 x 15 A, tail 20 A
(d) Head 10 x 12 A, tail 25 A

16. The substrate for photorespiration is

(a) Phosphoglyceric acid
(b) Glycolate
(c) Serine
(d) Glycine

17. Greatest producers of organic matter are

(a) Crop plants
(b) Forests
(c) Plants of the land area
(d) Phytoplankton of oceans

18. A very efficient converter of solar energy with net productivity of 204 kg/m^2 or more is the crop?

(a) Wheat
(b) Sugarcane
(c) Rice
(d) Bajra

19. A bottle filled with previously moistened mustard seeds and water was screw capped tightly and kept in a corner. It blew up suddenly after about half an hour. The phenomenon involved is

(a) Diffusion
(b) Imbibition
(c) Osmosis
(d) D.P.D

20. The principal pathway of water translocation in angiosperms is

(a) Sieve cells
(b) Sieve tube elements
(c) Xylem vessel system
(d) Xylem and phloem

21. Phototropic and geotropic movements are linked to

(a) Gibberellins
(b) Enzymes
(c) Auxin
(d) Cytokinins

22. Abscisic acid controls

(a) Cell division
(b) Leaf fall and dormancy
(c) Shoot elongation
(d) Cell elongation and wall formation

23. Phytohormones are

(a) Chemical regulation flowering

(b) Chemical regulation secondary growth

(c) Hormones regulating growth from seed to adulthood

(d) Regulators synthesised by plants and influencing physiological processes

24. A chemical believed to be involved in flowering is

(a) Gibberellin (b) Kinetin

(c) Florigen (d) Iba

25. The hormone responsible for apical dominance is

(a) IAA (b) GA

(c) ABA (d) Florigen

26. Twining of tendrils is due to

(a) Thigmotropism (b) Seismonasty

(c) Heliotropism (d) Diageotropism

27. Abscisic acid causes

(a) Stomatal closure (b) Leaf expansion

(c) Stem elongation (d) Root elongation

28. Photosynthetic pigments found in the chloroplasts occur in

(a) Thylakoid membranes (b) Plastoglobules

(c) Matrix (d) Chloroplast envelope

29. During monsoon, the rice crop of eastern states of India shows lesser yield due to limiting factor of

(a) C02 (b) Light

(c) Temperature (d) Water

30. With an increase in the turgidity of a cell, the wall pressure will be

(a) Fluctuate (b) Remain unchanged

(c) Increase (d) Decrease

31. When a cell is fully turgid, which of the following will be zero?

(a) Turgor pressure (b) Water potential

(c) Wall pressure (d) Osmotic pressure

32. NADPH2 is generated through

(a) Photosystem II
(b) Anaerobic respiration
(c) Glycolysis
(d) Photosystem I

33. A pigment which absorbs red and far-red light is

(a) Cytochrome
(b) Xanthophyll
(c) Phytochrome
(d) Carotene

34. The core metal of chlorophyll is

(a) Ni
(b) Cu
(c) Fe
(d) Mg

35. Chlorophyll-a molecule at its carbon atom 3 of the pyrrole ring II has one of the following

(a) Carboxylic group
(b) Magnesium
(c) Aldehyde group
(d) Methyl group

36. The water potential and osmotic potential of pure

(a) 100 and 200
(b) Zero and 100
(c) 100 and zero
(d) Zero and zero

37. Which combination of gases is suitable for fruit ripening?

(a) 80 per cent CH_4 and 20 per cent CO_2
(b) 80 per cent CO_2 and 20 per cent O_2
(c) 80 per cent C_2H_4 and 20 per cent CO_2
(d) 80 per cent CO_2 and 20 per cent CH_2

38. A plant hormone used for inducing morphogenesis in plant tissue culture is

(a) Cytokinins
(b) Ethylene
(c) Abscisic Acid
(d) Gibberellins

39. The rate of photosynthesis is higher in

(a) Very high light
(b) Continuous light
(c) Red light
(d) Green light

40. Main function of lenticel is

(a) Transpiration
(b) Guttation
(c) Gaseous exchange
(d) Bleeding

41. Choose the correct match Bladderwort, sundew, venus flytrap:

(a) Nepenthes, dionea, drosera

(b) Nepenthes, utricularia, vanda

(c) Utricularia, drosera, dionea

(d) Dionea, trapa, vanda

42. Cellular totipotency is demonstrated by

(a) Only gymnosperm cells (b) All plant cells

(c) All eukaryotic cells (d) Only bacterial cells

43. In soil, water available for plants is

(a) Gravitational water (b) Chemically bound water

(c) Capillary water (d) Hygroscopic water

44. The first step for initiation of photosynthesis will be

(a) Photolysis of water

(b) Excitement of chlorophyll molecules due to absorption of light

(c) Atp formation

(d) Glucose formation

45. Movement of leaves of sensitive plant, Mimosa pudica are due to

(a) Thermonasty (b) Seismonasty

(c) Hydrotropism (d) Chemically

46. Which country recently passed 'Sunshine Protection Act' making daylight saving time (DST) permanent?

(a) Russia (b) USA

(c) Brazil (d) Canada

47. Which Bank has recently launched 'Green Deposit Programme' for corporate clients?

(a) HDFC Bank (b) DBS India

(c) RBL Bank (d) HSBC India

48. MV Ram Prasad Bismil' recently became the longest vessel ever to sail on which river?

(a) Ganges (b) Brahmaputra

(c) Narmada (d) Krishna

49. Constitution (Scheduled Tribes) Order (Amendment) Bill was recently presented to include certain tribes in which state?

(a) Assam
(b) Madhya Pradesh
(c) Tripura
(d) Gujarat

50. Which state government recently unveiled Sports Policy 2022-27 to set up High Performance Centre (HPC)?

(a) Tamil Nadu
(b) Gujarat
(c) Rajasthan
(d) Punjab

ANSWER KEY

1	2	3	4	5	6	7	8	9	10	11	12	13	14	15
(c)	(a)	(c)	(c)	(b)	(b)	(a)	(c)	(d)	(d)	(d)	(d)	(c)	(a)	(c)
16	17	18	19	20	21	22	23	24	25	26	27	28	29	30
(b)	(d)	(b)	(b)	(c)	(c)	(b)	(d)	(c)	(a)	(a)	(a)	(a)	(b)	(c)
31	32	33	34	35	36	37	38	39	40	41	42	43	44	45
(b)	(d)	(c)	(d)	(d)	(d)	(c)	(a)	(c)	(c)	(c)	(b)	(c)	(b)	(b)
46	47	48	49	50										
(b)	(b)	(b)	(c)	(b)										

Chapter 33

C3, C4, CAM Pathways and their Differences, Respiration, Glycolate Pathway, Mechanism of Respiration, Respiration Quotient, R.Q. Values, Photoperiodism, Short, Long and Day Neutral Plants

1. **Which plants Scotoactive stomata perform**
   (a) C3 photosynthesis
   (b) C4 photosynthesis
   (c) Both (a) and (b)
   (d) CAM photosynthesis

2. **Stomata of CAM plants**
   (a) Are always open
   (b) Open during the day and close at night
   (c) Open during the night and close during the day
   (d) Never open

3. **Which element is located at the centre of the porphyrin ring in chlorophyll?**
   (a) Calcium
   (b) Magnesium
   (c) Potassium
   (d) Manganese

4. **The correct sequence of electron acceptor in ATP synthesis is**
   (a) Cyt. b, c, a3, a
   (b) Cyt. c, b, a, a3
   (c) Cyt. o, a, b, c
   (d) Cyt. b, c, a, a3

5. In C4 plants, CO_2 combines with

(a) Phosphoenol pyruvate
(b) Phosphoglyceraldehyde
(c) Phosphoglyceric acid
(d) Ribulose diphosphate

6. In C4 plants, CO_2 fixation is done by

(a) Sclerenchyma
(b) Chlorenchyma and hypodermis
(c) Mesophyll cells
(d) Guard cells

7. The primary acceptor, during CO_2 fixation in C3 paints, is

(a) Phosphoenolpyruvate (PEP)
(b) Ribulose 1, 5-diphosphate (rudp)
(c) Phosphoglyceric acid (PGA)
(d) Ribulose monophosphate (RMP)

8. The carbon dioxide acceptor in Calvin cycle/C3-plants is

(a) Phospho-enol pyruvate (PEP)
(b) Ribulose 1, 5-diphosphate (rubp)
(c) Phosphoglyceric acid (PGA)
(d) Ribulose monophosphate (RMP)

9. Which one is a C4-plant?

(a) Papaya
(b) Pea
(c) Potato
(d) Maize/corn

10. In C4 plants, Calvin cycle operates in

(a) Stroma of bundle sheath chloroplasts
(b) Grana of bundle sheath chloroplasts
(c) Grana of mesophyll chloroplasts
(d) Stroma of mesophyll chloroplasts

11. Kranz anatomy is typical of

(a) C4 plants
(b) C3 plants
(c) C2 plants
(d) CAM plants

12. A pigment which absorbs red and far-red light is

(a) Cytochrome
(b) Xanthophyll
(c) Phytochrome
(d) Carotene

13. Element present in nucleic acids, Phospholipids, ATP, ADP and NADPH is ______

(a) P (b) Na
(c) Bo (d) Co

14. ______ are required for chlorophyll synthesis

(a) Fe and Mg (b) Fe and Co
(c) Cu and Ca (d) Mg and Ca

15. In which form Ca is present in plants?

(a) Calcium pectate (b) Calcium carbonate
(c) Calcium (d) CaCl

16. Which element is essential for activating the enzyme but not a part of enzyme?

(a) Mn (b) Mg
(c) K (d) Co

17. For the selective permeability of the cell membranes which element is involved?

(a) K (b) Fe
(c) P (d) Ca

18. Which element is essential for pollen germination?

(a) Na (b) B
(c) Mo (d) Co

19. In photosynthesis which element is essential for the evolution of oxygen?

(a) CI (b) Mg
(c) Fe (d) All of the above

20. In leguminous plants leghemoglobin protects ______ activity

(a) Nitrogenase (b) Protiase
(c) Nitrate reductase (d) Both (a) and (b)

21. In Nitrogen fixation and metabolism which element is play major role?

(a) Mn (b) B
(c) Mg (d) Mo

22. Nitrobacter convert NO_2 into NO_3 in which process

(a) Reduction (b) Oxidation
(c) Both (a) and (b) (d) None of the above

23. In plant ______ is responsible for guttation

(a) Root pressure (b) Photosynthesis
(c) Transpiration (d) Osmosis

24. Which is selectively permeable membrane of the cell ______

(a) Cell wall (b) Cytoplasm
(c) Plasmalemma (d) None of the above

25. The chemical use an antitranspirant is ______

(a) Salicylic acid (b) Abscisic acid
(c) Phenylmercuric acid (d) All of the above

26. During high wind velocity, the stomata of plants ______

(a) Remain open (b) Close down
(c) Open more widely (d) Remain unaffected

27. Selective passage across a semipermeable membrane is called ______

(a) Osmosis (b) Diffusion
(c) Both (a) and (b) (d) Imbibition

28. Stomata open at night but close during day time in which plants?

(a) Xerophytes (b) Mesophytes
(c) Succulents (d) Both (a) and (b)

29. Which enzyme plays major role in opening and closing of stomata?

(a) Beta-amylase (b) Pyruvic kinase
(c) RuDP (d) PEP carboxylase

30. Which nutrient is obtained by insectivorous plants from insect bodies?

(a) P (b) K
(c) N (d) Na

31. In Himalayan Forest commonly N2 fixing bacteria is ______

(a) Azotobacter (b) Frankia
(c) Nitrobacter (d) None of the above

32. The enzyme responsible for atmospheric nitrogen fixation is ______

(a) Nitrogenase (b) Amylase
(c) Nitrogenase and Amylase (d) Lemda nitrogenage

33. In which method ions are absorbed by plants

(a) By carriers and pumps (b) By carriers method
(c) By difference in DPD (d) By molecular diffusion

34. Bidirectional translocation of minerals occurs in plants

(a) Phloem (b) Xylem
(c) Cambium (d) Perenchyma and phloem

35. Nitrification process is done by ______

(a) *Rhizobium* (b) *B. bassiana*
(c) *Pseudomonas* (d) *Nitrococcus*

36. Transpiration mostly affected by ______

(a) Humidity (b) Temperature
(c) Light (d) Wind

37. Which enzyme is not proteinaceous in nature?

(a) Ribozyme (b) Both (a) and (b)
(c) Isozyme (d) Trypsin

38. Pyruvate dehydrogenease complex and glycolytic pathway is located in ______

(a) Cytosol and Mitochondria (b) Cytosol
(c) Golgi bodies (d) Microsomes

39. Most C4 plants are native ______

(a) Xerix habitats (b) Merine habitats
(c) Grass habitats (d) All habitats

40. Which is involve in photosynthesis

(a) Blackman and Hill (b) Krebs cycle reaction
(c) E.M.P. Pathway (d) None of the above

41. Hill reaction take place in Granna formed

(a) ATP and NADPH2 both (b) NADPH2
(c) ADP (d) All of the above

42. Anabolic process is ______

(a) Photosynthesis (b) Transpiration
(c) Both (a) and (b) (d) None of the above

43. Absorption of light take place during photosynthesis

(a) Photosystem-I only (b) Photosystem-II only
(c) Photosystem-I and Photosystem-II (d) None of the above

44. In C4 pathway

(a) O_2 combine with PGA
(b) CO_2 combines with PEP
(c) CO_2 combine with RMP
(d) All of the above

45. Dark reaction of photosynthesis occurs in ______

(a) Stomata
(b) Stroma and matrix
(c) Stroma
(d) Granna

46. The 'Yilan' Crater, the largest impact crater on Earth, has been discovered in which country?

(a) Russia
(b) China
(c) Australia
(d) Japan

47. "Business Blasters Investment Summit and Expo" is an initiative of which state/UT Government

(a) Telangana
(b) New Delhi
(c) Tamil Nadu
(d) Maharashtra

48. Which institution released the 'State of India's Environment Report 2022'?

(a) NITI Aayog
(b) Centre for Science and Environment
(c) Ministry of Environment, Forest and Climate Change
(d) UNDP

49. Which Asian country has recently imposed its longest power cut since 1996?

(a) Bangladesh
(b) Indonesia
(c) Sri Lanka
(d) Afghanistan

50. The 'Sagar Praikrama' Program was inaugurated at which state?

(a) Kerala
(b) Gujarat
(c) Tamil Nadu
(d) Maharashtra

ANSWER KEY

1	2	3	4	5	6	7	8	9	10	11	12	13	14	15
(d)	(c)	(b)	(d)	(a)	(c)	(b)	(b)	(d)	(a)	(a)	(c)	(a)	(a)	(a)
16	17	18	19	20	21	22	23	24	25	26	27	28	29	30
(c)	(d)	(b)	(a)	(c)	(d)	(b)	(a)	(c)	(d)	(b)	(a)	(c)	(d)	(c)
31	32	33	34	35	36	37	38	39	40	41	42	43	44	45
(b)	(a)	(a)	(a)	(a)	(a)	(a)	(a)	(a)	(a)	(a)	(a)	(c)	(b)	(c)
46	47	48	49	50										
(b)	(b)	(b)	(c)	(b)										

STATISTICS

Chapter 34

Statistics, Central Tendency, Dispersion Measure, Distributions, Correlation, Regression, Degree of Freedom, Hypothesis, Chi-square Test, Z Test, t Test, ANOVA, Experimental Designs, Split Plot Technique, Strip Plot Technique

1. **Local control is not applied for the design**
   (a) RBD (b) CRD
   (c) LSD (d) SPD
2. **For an asymmetrical data if mean=8 and median 6 then value of mode is**
   (a) 2 (b) 8
   (c) 3 (d) 12
3. **2nd central moment is equal to**
   (a) Mean (b) Skewness
   (c) Kurtosis (d) SD
4. **For symmetrical data value of kurtosis is**
   (a) >3 (b) 3
   (c) <3 (d) None of the above
5. **Which one of the following is not an absolute measure of dispersion**
   (a) Mean deviation (b) SD
   (c) Variance (d) Coefficient of variation

6. ANOVA was defined by

(a) Pearson (b) Fisher

(c) Secrist (d) Bowly

7. Central tendency suitable for size of agricultural holdings

(a) Mean (b) Median

(c) Mode (d) All of the above

8. Chi square test was given by

(a) Fisher (b) Karl pearson

(c) gosset (d) De moivre

9. 2nd quartile is equal to

(a) Mid-point of the data

(b) Mid-point of the first half of the data

(c) Mid-point of the second half of the data

(d) None of the above

10. Father of vital statistics

(a) John grant (b) Fisher

(c) Smith (d) None of the above

11. Range of probability

(a) 0 to 1 (b) 1 to 1

(c) -1 to +1 (d) None of the above

12. Probability of impossible event is always

(a) 0 (b) 1

(c) 0 to 1 (d) None of the above

13. Regression coefficient lies between

(a) $-\infty$ to $+\infty$ (b) -1 to 1

(c) 0 to 1 (d) None of the above

14. Correlation coefficient lies between

(a) -1 to +1 (b) 0 to 1

(c) $-\infty$ to $+\infty$ (d) None of the above

15. The value of chi-square always lies between

(a) 0 to ∞ (b) -1 to +1

(c) 0 to +1 (d) None of the above

16. The value of standard deviation between

(a) 0 to ∞ (b) -1 to +1

(c) 0-1 (d) None of the above

17. In the case of perfect negative correlation, the coefficient of correlation will be

(a) -1 (b) 0

(c) 1 (d) -1 to 0

18. Value of 0th central moment is

(a) 0 (b) 1

(c) equal to mean (d) varies

19. Term regression

(a) Fisher (b) Karl Pearson

(c) Francis Galton (d) Captain John Grant

20. Error degree of freedom for 6 replications in LSD.

(a) 36 (b) 25

(c) 20 (d) 12

21. Goodness of test is used for distribution of

(a) Chi square test (b) Z test

(c) t test (d) F test

22. To find the average height of plants we should use

(a) Arithmetic mean (b) Geometric mean

(c) Harmonic mean (d) None of the above

23. When sample size is small and S.D is unknown which test is used

(a) F test (b) Z test

(c) t test (d) Chi square test

24. For testing significance of correlation coefficient use

(a) F test (b) Z test

(c) t test (d) None of the above

25. Most commonly used central tendency is

(a) Arithmetic mean (b) Median

(c) Mode (d) Geometric mean

26. To get higher precision is required

(a) Larger samples
(b) Smaller sample
(c) One sample
(d) None of the above

27. The minimum sample size for using chi square test is

(a) 5
(b) 25
(c) 50
(d) 75

28. The student t-test was discovered by

(a) W.S Gosset
(b) C.J. Grant
(c) R.A Fisher
(d) None of the above

29. To find the average size of shoes sold in the market we should use

(a) Mean
(b) Median
(c) Mode
(d) None of the above

30. To present the whole data by only single value we used

(a) Measure of Dispersion
(b) Measure of central tendency
(c) Measure of Skewness
(d) Measure of kurtosis

31. During the experimentation, some lost information can be get by

(a) Missing plot technique
(b) Strip plot technique
(c) Split plot technique
(d) None of the above

32. To study effect of two or more factor require different plot size

(a) Split plot technique
(b) Strip plot technique
(c) Missing plot technique
(d) None of the above

33. If all the variate values are negative the standard deviation will be

(a) Negative
(b) positive
(c) zero
(d) None of the above

34. The design where the number of replications are equal to treatments

(a) CRD
(b) LSD
(c) RBD
(d) Split plot technique

35. When the two variables moves in a opposite direction, correlation is said to be

(a) Positive
(b) Negative
(c) Zero
(d) None of the above

36. When the two variables moves in same direction, Correlation is said to be

(a) Positive (b) Negative

(c) Zero (d) None of the above

37. The number of independent constraints in the set of observation is called

(a) Skewness (b) Kurtosis

(c) Degree of freedom (d) None of the above

38. Which of the following is the Relative measure of dispersion

(a) Range (b) mean deviation

(c) Standard deviation (d) Coefficient of variation

39. Simplest measure of dispersion is

(a) Range (b) Mean deviation

(c) Standard deviation (d) Coefficient of variaton

40. Most commonly used measure of dispersion is

(a) Range (b) Mean deviation

(c) Standard deviation (d) Variance

41. Square of Standard deviation is known as

(a) C.V (b) Variance

(c) Mean deviation (d) range

42. Which experimental design is used in Pot experiment

(a) CRD (b) RBD

(c) LSD (d) None of the above

43. Which experimental design is used in field experiments

(a) LSD (b) CRD

(c) RBD (d) None of the above

44. One way classification and no way control is seen in which experimental design

(a) CRD (b) LSD

(c) RBD (d) None of the above

45. Two way classification and one way control is observed in

(a) CRD (b) RBD

(c) LSD (d) None of the above

46. Z test is used when the sampole size is more than

(a) 30 (b) 20

(c) 10 (d) 5

47. t test is used when the sample size is

(a) > 30 (b) < 30

(c) Any sample (d) None of the above

48. Basic principles of field experimentation are

(a) Replication (b) Randomization

(c) Local control (d) All of the above

49. Rejecting null hypothesis when it is true comes under which type of error

(a) Type 1 error (b) Type 2 error

(c) Type 3 error (d) None of the above

50. Accepting null hypothyesis when it is false comes under

(a) Type 1 error (b) Type 2 error

(c) Type 3 error (d) None of the above

ANSWER KEY

1	2	3	4	5	6	7	8	9	10	11	12	13	14	15
(b)	(a)	(d)	(b)	(d)	(b)	(c)	(b)	(a)	(a)	(a)	(a)	(a)	(a)	(a)
16	17	18	19	20	21	22	23	24	25	26	27	28	29	30
(a)	(a)	(b)	(c)	(c)	(a)	(a)	(c)	(c)	(a)	(a)	(c)	(a)	(c)	(b)
31	32	33	34	35	36	37	38	39	40	41	42	43	44	45
(a)	(a)	(b)	(b)	(b)	(a)	(c)	(d)	(a)	(c)	(b)	(a)	(c)	(a)	(b)
46	47	48	49	50										
(a)	(b)	(d)	(a)	(b)										

ANIMAL HUSBANDRY

Chapter 35

Animal Production

1. **Milk of Which breed has lowest fat percent**
   (a) Haryana
   (b) Sahiwal
   (c) Holstien
   (d) Jersey

2. **Whey is produced from**
   (a) Heating colostrum
   (b) After milk coagulation
   (c) Butter production
   (d) Cheese production

3. **Docking means removal of**
   (a) Tail
   (b) Hair
   (c) Horn
   (d) Hoof

4. **Pashmina wool obtained from**
   (a) Yak
   (b) Cattle
   (c) Sheep
   (d) Goat

5. **Buffer value of milk is due to**
   (a) Protein
   (b) Fat
   (c) Lipid
   (d) Carbohydrate

6. **Q fever through milk is caused by**
   (a) Virus
   (b) Bacteria
   (c) Mycoplasma
   (d) Protozoa

7. **Disbudding of newborn cow is done in**
   (a) 4-10 days (b) 15-20 days
   (c) 1 month (d) Within 3 months

8. **Milk fat is measured by**
   (a) Lactometer (b) Hortvet cryoscope
   (c) Butyrometer (d) Visicometer

9. **Maximum overrun of butter is**
   (a) 16 per cent (b) 25 per cent
   (c) 36 per cent (d) 45 per cent

10. **Fat percent in milk of buffalo**
    (a) 3.5 (b) 5.2
    (c) 6.6 (d) 9

11. **Protein per cent in dahi**
    (a) 3.2 - 3.4 per cent (b) 4.2- 5.2 per cent
    (c) 1.2- 2.3 per cent (d) None of the above

12. **Fat per cent in colustrum of sheep**
    (a) 4.5 per cent (b) 3.6 per cent
    (c) 5.6 per cent (d) 7.4 per cent

13. **Disease found in sheep during monsoon**
    (a) Foot rot (b) FMD
    (c) Rinderpest (d) All of the above

14. **Most dangerous disease of chicken?**
    (a) chicken pox (b) white diarrhoea
    (c) Mark's disease (d) Ranikhet

15. **Spleenic fever is common name for**
    (a) Anthrax (b) Rinderpest
    (c) FMd (d) Ranikhet

16. **Lincoln is a breed of**
    (a) Australia (b) Switzerland
    (c) England (d) France

17. Merino of Rajasthan

(a) Bikaneri (b) Chokla

(c) Marwari (d) Jaisalmeri

18. Throwing animals on ground is known as

(a) Culling (b) Docking

(c) Flushing (d) Casting

19. Most variable constituent of milk is

(a) Fat (b) Proteins

(c) Cho (d) Casein

20. Oxidation of fat is

(a) Beta oxidation (b) Alpha oxidation

(c) Both (d) None of the above

21. Meat of goat is known as

(a) Chevon (b) Mutton

(c) Pork (d) Buffin

22. Freshly drawn milk has pH of

(a) 6.5-6.7 (b) 5.5-5.7

(c) 4.6-4.8 (d) None of the above

23. The poultry specially raised for meat production is

(a) Broilers (b) Layers

(c) Both (d) None of the above

24. Largest sheep breed

(a) Lohi (b) Nellore

(c) Marwari (d) Shahabadi

25. Tallest sheep breed

(a) Nellore (b) Mandya

(c) Lohi (d) Chokla

26. Exotic breed of dairy goat known as Jersey cow

(a) Alpine (b) Saanen

(c) Toggenberg (d) Anglo nubian

27. Chocolate coloured breed of goat

(a) Toggenberg (b) Alpine

(c) Angora (d) Beetul

28. Excellent breed of goat for meat

(a) Black bengal (b) Beetul

(c) Marwari (d) Sirohi

29. Dairy breed example is

(a) Sahiwal (b) Malvi

(c) Nagori (d) Amaratmahal

30. Dahil native of

(a) Sri Lanka (b) India

(c) Bangladesh (d) Pakistan

31. Gir native of

(a) Orissa (b) Gujarat

(c) Jammu (d) Tamil Nadu

32. Sindhi native of

(a) Karachi (b) Bangladesh

(c) Afganisthan (d) India

33. Superior among drought breed

(a) Malvi (b) Gir

(c) Nagori (d) Sahiwal

34. Dual purpose breed

(a) Haryana (b) Kan kreji

(c) Both (d) None of the above

35. Highest body weight in Indian cows

(a) Gir (b) Malvi

(c) Rathi (d) Kan krej

36. Best dual-purpose breed of cow

(a) Haryana (b) Rathi

(c) Malvi (d) Gir

37. Copper colour body coat breed

(a) Zaffabadi (b) Godavari

(c) Murrah (d) Bhadwari breed

38. Sickle shape horn is present in

(a) Surati (b) Murrah

(c) Godavari (d) Nil ravi

39. Milk is white in colour due to

(a) Casein
(b) Carontene
(c) Both
(d) None of the above

40. Milk is poor source of

(a) Vit-C
(b) Fe
(c) Both
(d) None of the above

41. Crazy chick disease caused due to deficiency of

(a) Vit-k
(b) Vit-E
(c) Vit-D
(d) Vit-c

42. Mating of the related individual with 4-6 generation is known as

(a) Inbreeding
(b) Linebreeding
(c) Close breeding
(d) Outcrossing

43. World's No.1 egg producer

(a) New Hempshire
(b) Leghorn
(c) English class
(d) Rhode Island Red

44. Horse is a

(a) Ruminant hoofed
(b) Non ruminant hoofed
(c) Even toed hoofed
(d) Odd toed hoofed

45. Methylene Blue Reduction Test (MBRT) is conducted to know the efficiency of

(a) Toned milk
(b) Double toned milk
(c) Sterilized milk
(d) Pasteurized milk

46. When two breeds are crossed alternatively is known as

(a) Criss crossing
(b) Triple crossing
(c) Back crossing
(d) None of the above

47. The best age of dehorning is

(a) 1 week
(b) 4 weeks
(c) 8 weeks
(d) 16 weeks

48. Better quality khoa is prepared from

(a) Buffalo milk
(b) Cow milk
(c) Sheep milk
(d) Camel milk

49. Marek's disease found in

(a) Cattle
(b) Poultry
(c) Sheep
(d) Goat

50. Yellow colour of egg the is due to–

(a) Carotene
(b) Anthocyanin
(c) Vitamin B
(d) Xanthophyll

ANSWER KEY

1	2	3	4	5	6	7	8	9	10	11	12	13	14	15
(c)	(d)	(a)	(d)	(a)	(c)	(a)	(a)	(b)	(c)	(a)	(d)	(a)	(d)	(a)
16	17	18	19	20	21	22	23	24	25	26	27	28	29	30
(b)	(b)	(d)	(a)	(a)	(a)	(a)	(a)	(a)	(a)	(d)	(a)	(a)	(a)	(d)
31	32	33	34	35	36	37	38	39	40	41	42	43	44	45
(b)	(a)	(c)	(c)	(d)	(a)	(d)	(a)	(a)	(c)	(b)	(a)	(b)	(b)	(d)
46	47	48	49	50										
(a)	(a)	(a)	(b)	(d)										

GRAND TESTS FOR ALL THE SUBJECTS (10 TESTS)

Grand Test-1

1. **Truck gardening word**
   (a) Latin
   (b) Greek
   (c) French
   (d) Spanish
2. **Monsoon word**
   (a) Latin
   (b) Greek
   (c) Arabic
   (d) French
3. **Elisa test is used to detect**
   (a) Fungi
   (b) Bacteria
   (c) Virus
   (d) MLO
4. **Ooze test is used to detect**
   (a) Fungi
   (b) Bacteria
   (c) Virus
   (d) MLO
5. **Most of plant viruses contains**
   (a) Ss RNA
   (b) Ds RNA
   (c) Ss DNA
   (d) None of the above
6. **Cauliflower mosaic virus contains**
   (a) DNA
   (b) RNA
   (c) Either of the nuclic acid
   (d) None of the above

7. Father of microbiology

(a) Louies pasture (b) Smith
(c) Antony debary (d) None of the above

8. Father of plant pathology

(a) Antony debary (b) Smith
(c) Luies pasture (d) None of the above

9. Head quarter of directorate of plant protection quarantine and storage

(a) Haryana (b) New delhi
(c) Madhya Pradesh (d) Jarkhand

10. Protien hydrolysate is used as attractent for controlling of

(a) Pumpkin beetle (b) Fruit fly
(c) Chilli borer (d) None of the above

11. Tree banding is a control measure for

(a) Mango mealy bug (b) Citrus mealy bug
(c) Both (d) None of the above

12. Bud necrosis in musk melon transmit by

(a) Thrips (b) Aphids
(c) White fly (d) None of the above

13. Bud necrosis in ground nut transmit by

(a) White fly (b) Aphid
(c) Thrips (d) None of the above

14. Alphonso var of mango grown in which state

(a) Uttar Pradesh (b) Madhya Pradesh
(c) Maharashtra (d) Andhra Pradesh

15. California papershell var of

(a) Walnut (b) Almond
(c) Pecanut (d) Pomegranate

16. Pajaro is a var of

(a) Kiwi (b) Strawberry
(c) Cherry (d) None of the above

17. In which country 1st zero tillage introduced

(a) India (b) USA
(c) Brazil (d) Isreal

18. Black cotton soils deficient in

(a) Zn (b) N

(c) P (d) B

19. In black soil dominant clay mineral

(a) Montmorlite (b) Kaolinite

(c) Both (d) None of the above

20. In Red soils dominant clay mineral

(a) Kaolinite (b) Montmorlite

(c) Both (d) None of the above

21. In addition of organic matter bulk density

(a) Increase (b) Decrease

(c) Stable (d) None of the above

22. Crops takes ammonical form directly

(a) Paddy (b) Potato

(c) Both (d) None of the above

23. According to Icar soil is classified into

(a) 4 (b) 6

(c) 8 (d) 10

24. Chlorites type of mineral

(a) 2:1:1 (b) 2:1

(c) 1:1 (d) None of the above

25. Vermiculite type of mineral?

(a) 2:1:1 expanding (b) 2:1 expanding

(c) 2:1 limited expanding (d) None of the above

26. Montmorlite type of mineral

(a) 2:1 expanding (b) 2:1 non expanding

(c) 1:1 expanding (d) 2:1:1 expanding

27. Kaolinite type of mineral

(a) 1:1 expanding (b) 1:1 non expanding

(c) 2:1 expanding (d) 2:1 non expanding

28. Bulk density of soil

(a) 2.65 g/cc (b) 1.33 g/cc

(c) Both (d) None of the above

29. Particle density of soil

(a) 2.65 g/cc (b) 1.33 g/cc

(c) Both (d) None of the above

30. Clay particle size

(a) <0.002mm (b) <0.02mm

(c) <0.02cm (d) <0.2mm

31. Process of mixing of soil

(a) Pedoturbation (b) Podzolization

(c) Both (d) None of the above

32. Process of moving out of sesiquioxide is known as

(a) Podzolization (b) Pedoturbation

(c) Both (d) None of the above

33. Canker nodules are found in

(a) Black soils (b) Red soils

(c) Both (d) None of the above

34. Biofertilizer suitable for wheat and sugarcane

(a) Azatobacter (b) Azosporillum

(c) Rhizobium (d) None of the above

35. Vertical mulch used for soil

(a) Black cotton soils (b) Red soil

(c) Both (d) None of the above

36. Reclamation disease due to deficiency of

(a) Fe (b) Zn

(c) B (d) Cu

37. Plants grown in salt water

(a) Saltophytes (b) Halophytes

(c) Epiphytes (d) None of the above

38. Jhum cultivation

(a) Northern parts of India (b) Western part

(c) Eastern part (d) Southern part

39. Micro nutrient deficiency in Indian soils

(a) Zn (b) Fe

(c) B (d) Cu

40. Amino acid deficient in legumes

(a) Lysine (b) Methionin
(c) Both (d) None of the above

41. Amino acid deficient in cereals

(a) Lysin (b) Methionine
(c) Both (d) None of the above

42. Maximum area under irrigation

(a) Uttar Pradesh (b) Andhra Pradesh
(c) Punjab (d) Tamil Nadu

43. Forest conservation act

(a) 1908 (b) 1980
(c) 1890 (d) 1809

44. Highest area under drip irrigation recent data?

(a) Maharashtra (b) Andhra Pradesh
(c) Tamil Nadu (d) Punjab

45. Highest fertilizer consumption state

(a) Uttar Pradesh (b) Punjab
(c) Andhra Pradesh (d) Madhya Pradesh

46. Average rain fall of India

(a) 1194 mm (b) 1149 cm
(c) 1914 mm (d) 1194 cm

47. 1st state in food grain production

(a) Madhya Pradesh (b) Uttar Pradesh
(c) Andhra Pradesh (d) Punjab

48. Gross cropped area and Net cropped area of India?

(a) 193mha, 141mha (b) 139mha, 134 mha
(c) 193mha, 134 mha (d) 191mha,143 mha

49. Soil science word

(a) Latin (b) Greek
(c) French (d) Arabic

50. Cell word

(a) Latin (b) Greek
(c) French (d) Arabic

ANSWER KEY

1	2	3	4	5	6	7	8	9	10	11	12	13	14	15
(c)	(c)	(c)	(b)	(a)	(a)	(a)	(a)	(a)	(b)	(a)	(b)	(c)	(c)	(b)
16	17	18	19	20	21	22	23	24	25	26	27	28	29	30
(b)	(b)	(b)	(a)	(a)	(b)	(c)	(c)	(a)	(c)	(a)	(b)	(b)	(a)	(a)
31	32	33	34	35	36	37	38	39	40	41	42	43	44	45
(a)	(a)	(b)	(a)	(a)	(d)	(b)	(c)	(a)	(b)	(a)	(c)	(b)	(b)	(b)
46	47	48	49	50										
(a)	(b)	(a)	(a)	(a)										

Grand Test 2

1. **Transpiration pull theory**
   (a) Dixon and jolly (b) Jemson
   (c) Hales (d) Boehm
2. **Root hairs are absent in hydrophytes**
   (a) True (b) False
3. **Pulsation theory given by sir J.C Bose**
   (a) True (b) False
4. **Theory of profit**
   (a) Walker (b) Malthus
   (c) Lerner (d) Adam smith
5. **Modern theory of interest**
   (a) Walker (b) Hicks-Hansen
   (c) Malthus (d) Adam smith
6. **Population theory**
   (a) Walker (b) Malthus
   (c) Lerner (d) Adam smith
7. **Theory of inflation**
   (a) Walker (b) Malthus
   (c) Lerner (d) Adam smith

8. Theory of absolute advantage

(a) A walker (b) Malthus

(c) Lerner (d) Adam smith

9. Marginal farmer

(a) <1 ha (b) 1-2 ha

(c) 2-4 ha (d) >4 ha

10. Small farmer

(a) <1 ha (b) 1-2 ha

(c) 2-4 ha (d) >4 ha

11. Medium farmer

(a) <1 ha (b) 1-2 ha

(c) 2-4 ha (d) >4 ha

12. Large farmer

(a) <1 ha (b) 1-2 ha

(c) 2-4 ha (d) >4 ha

13. Element related to pollen tube germination

(a) Na (b) B

(c) Mg (d) Fe

14. Onion pollination is done by

(a) Honey bees (b) Wind

(c) Hydrophyly (d) Ornithophyly

15. Buttoning is due to def of

(a) N (b) B

(c) Ca (d) Mg

16. Pagoda related to

(a) Mahal garden (b) Japanese garden

(c) Italian garden (d) English garden

17. In onion bolting takes place due to high temperature

(a) True (b) False

18. Pusa baisakhi is a variety of

(a) Moong bean (b) Rice

(c) Wheat (d) Paddy

19. Depth of water required by a crop is known as delta

(a) True (b) False

20. Bio climatic low proposed by

(a) Watson (b) T.N. knight

(c) Jemson (d) Hopkins

21. Rancidity in sunflower oil is due to

(a) Oxidation (b) Reduction

(c) Both (d) None of the above

22. Crop with highest K Uptake

(a) Tomato (b) Potato

(c) Bhendi (d) Brinjal

23. Erosion permitting crop

(a) Rice (b) Wheat

(c) Maize (d) Jower

24. 1st AICRP started in which crop

(a) Rice (b) Wheat

(c) Maize (d) Jower

25. Sickle shape horn

(a) Surati (b) Murrah

(c) Godavari (d) Nil ravi

26. Copper colour body coat breed

(a) Zaffabadi (b) Godavari

(c) Murrah (d) Bhadwari breed

27. Value which occur most frequently

(a) Mean (b) Median

(c) Mode (d) Range

28. Simplest measure of dispersion

(a) Mean (b) Median

(c) Mode (d) Range

29. _______used to find average speed

(a) Harmonic mean (b) Mode

(c) Range (d) Arithmetic mean

30. Used for qualitative data

(a) Mean (b) Median

(c) Mode (d) Range

31. Used for meteorological forecasting

(a) Mean (b) Median

(c) Mode (d) Range

32. Harmonic mean is reciprocal of arthamatic mean

(a) True (b) False

33. Standard deviation given by Karl Pearson

(a) True (b) False

34. _______most commonly used to measure dispersion

(a) Standard deviation (b) Root mean square deviation

(c) Both (d) None of the above

35. Term variance

(a) Fisher (b) Johenson

(c) Pearson (d) Bateson

36. Dairy breed

(a) Sahiwal (b) Malvi

(c) Nagori (d) Amaratmahal

37. Yield is a

(a) Continuous variable (b) Discontinuous variable

(c) Both (d) None of the above

38. Cooperative credit society act was passed in

(a) 1900 (b) 1902

(c) 1903 (d) 1904

39. The income elasticity of inferior goods is positive

(a) True (b) False

40. Absorption of water by dry woods-imbibition

(a) True (b) False

41. Eye drops are

(a) Hypotonic (b) Hypertonic

(c) Isotonic (d) None of the above

42. Term genetics

(a) Bateson (b) Vavilov

(c) Wolf (d) Darwin

43. Germplasm theory

(a) Weisman (b) Darwin

(c) Wolf (d) Sutton

44. Mutation 1st defined by ?

(a) Fairchaild (b) Vavilov

(c) Bateson (d) Hugo de vries

45. Chromosomal theory of inheritance

(a) Wolf (b) Lamark

(c) Weisman (d) Sutton and boveri

46. Centers of origin 1st given by ?

(a) Vavilov (b) Muller

(c) Johenson (d) Morgan

47. Term pureline

(a) Johansen (b) Morgan

(c) Muller (d) Vavilov

48. "fungi and plant diseases" book by ?

(a) Mundkur (b) Butler

(c) Kirtikar (d) Mehta

49. "Fungi and diseases in plants" book written by ?

(a) Kirtikar (b) Mehta

(c) Mundkur (d) Butler

50. Two kinds of resistance is suggested by?

(a) Vanderplank (b) Flor

(c) Burril (d) Brefeld

ANSWER KEY

1	2	3	4	5	6	7	8	9	10	11	12	13	14	15
(a)	(a)	(a)	(a)	(b)	(b)	(c)	(d)	(a)	(b)	(c)	(d)	(b)	(a)	(a)
16	17	18	19	20	21	22	23	24	25	26	27	28	29	30
(b)	(a)	(a)	(a)	(d)	(a)	(b)	(c)	(c)	(a)	(d)	(c)	(d)	(a)	(b)
31	32	33	34	35	36	37	38	39	40	41	42	43	44	45
(c)	(a)	(a)	(c)	(a)	(a)	(a)	(d)	(b)	(a)	(c)	(a)	(a)	(d)	(d)
46	47	48	49	50										
(a)	(a)	(a)	(d)	(a)										

Grand Test - 3

1. **DRIS approach for recommending fertilizer schedule based on**
   (a) Soil sample
   (b) Plant sample
   (c) Water sample
   (d) Both (a) and (b)
2. **Long term persistent herbicide is**
   (a) Pendimethalin
   (b) Paraquat
   (c) 2,4 D
   (d) Glyphosate
3. **Bioassay of herbicide 2,4**
   (a) Soybean
   (b) Sorghum
   (c) Sugarbeet
   (d) Cotton
4. **Cloud which resembles like mare tail is**
   (a) Cirrus
   (b) Altocumulus
   (c) Stratus
   (d) Nimbus
5. **Which one in field of PRH10 will be considered as rouge**
   (a) *Cyprus* spp.
   (b) *Echinocloa* spp.
   (c) Wild rice
   (d) Sambha mahsoori
6. **Genetically modified mustard was developed by**
   (a) Pro agro seed company ltd
   (b) Syngenta
   (c) Monsanto
   (d) Advanta

7. Khaira disease is found in

(a) Oilseeds and legumes (b) Pearl millet
(c) Maize (d) Rice

8. 2 4-D affects sedges but not grasses

(a) True (b) False

9. Tapping is done in

(a) Tea (b) Rubber
(c) Cotton (d) Coffee

10. Wolf apple is the common name of

(a) Potato (b) Tomato
(c) Pomato (d) Colocasia

11. Among various classes of salt affected soil,the most permeable one is

(a) Saline (b) Saline sodic
(c) Non saline sodic (d) Degraded sodic

12. Slaking of minerals is called

(a) Hydrogenation (b) Dehydrogenation
(c) Hydration (d) Hydroboration

13. Lightness or darkness of soil colour indicates

(a) Hue (b) Value
(c) Chroma (d) Soma

14. Value of bemmelen factor is

(a) 0.453 (b) 2.245
(c) 1.234 (d) 1.724

15. Thermal conductivity is highest in

(a) Sand (b) Silt
(c) Clay (d) Peat

16. Most recent soils are

(a) Inceptisol (b) Entisol
(c) Alfosol (d) Mollisol

17. Equivalent acidity of anhydrous ammonia is

(a) 128 (b) 46
(c) 98 (d) 148

18. End product of denitrification is

(a) N_2 (b) NO_3

(c) NO_2 and N_2 (d) NH_3

19. Concept of cardinal temperature was given by

(a) Sachs (b) Helmont

(c) Tisdale (d) Liebig

20. First fertilizer manufactured in India was

(a) DAP (b) SSP

(c) CAN (d) MOP

21. No. of essential amino acid is

(a) 5 (b) 10

(c) 15 (d) 17

22. Genetic constituent of female drosophila is

(a) XX (b) XY

(c) ZZ (d) ZW

23. Which of the following DNA marker is co-dominant in nature

(a) RFLP (b) AFLP

(c) RAPD (d) A and B

24. Haploids can be produced experimentally from

(a) Meristem (b) Anther

(c) Endosperm (d) Pith

25. Lecithine phospholipids synthesised in

(a) ER (b) Mitochondria

(c) Golgibody (d) Cytoplasm

26. Inbred lines can be improved by

(a) Transgressive breeding (b) Mutation breeding

(c) Back cross method (d) All of the above

27. Dead hard seed is found in

(a) Okra (b) Castor

(c) Coconut (d) Soyabean

28. Who said"Omnis cellula e cellula"

(a) Nageli (b) Schwan

(c) Robertson (d) Virchow

29. Which of the following is organelle within organell

(a) Ribosome (b) Nucleus
(c) Nucleolus (d) Chloroplast

30. Quantitative inheritance is through

(a) Effective genes (b) Polygenes
(c) Oligogenes (d) Ultra genes

31. Fungi was first studied by

(a) Kuhne (b) De bary
(c) Gauman (d) Michaeli

32. Non reducing sugar is

(a) Glucose (b) Sucrose
(c) Fructose (d) Lactose

33. When male nuclei migrate into oogonium through fertilization tube is called

(a) Spermatization (b) Gametangial copulation
(c) Gametangial contact (d) Parasexuality

34. Molya disease is caused by

(a) Physiological disorder (b) Bacteria
(c) Phytoplasma (d) Nematodes

35. Gram -ve bacteria is

(a) Acetobactor (b) Bacillus
(c) Clostridium (d) Actinomycetes

36. 1st stage of wheat rust is

(a) Pycnial (b) Aecial
(c) Uredial (d) Basidial

37. 1st commercially used organic fungicide was

(a) Bordeaux mixture (b) Elemental sulfer
(c) Spergon (d) Nabam

38. Blotter test is used for

(a) Seed health (b) Seed vigour
(c) Seed viability (d) Seed grading

39. Most suitable for control of early blight of potato

(a) Zineb (b) Dithane M 45

(c) Blitox 50 (d) Difolatan

40. Tea rust is caused by

(a) Algae (b) MLO

(c) PPLO (d) Viroids

41. Toxaphene is a

(a) Contact poison (b) Stomach poison

(c) Both (d) Pheromone

42. Pebrine disease caused in

(a) Honey bee (b) Lac insect

(c) Silkworm (d) Wasp

43. Raynia is effective against

(a) Coleoptera (b) Hemiptera

(c) Diptera (d) Lepidoptera

44. Lantana bug discovered by

(a) Chandra sekhar lohmi (b) Hem singh pruthi

(c) Yv singh (d) Bewley

45. Which one is the enemy of honey bee

(a) *Gangora thyrsis* (b) *Vesa cincta*

(c) *Atherigona soccata* (d) *Galleria mellonella*

46. Red cotton bug is

(a) *Dysdercus hoenigii* (b) *Sylepta derogata*

(c) *Chilo auricilius* (d) *Earis vittella*

47. Allomones favours receiver

(a) True (b) False

48. Parasite attacks on primary parasites are known as

(a) Secondary parasite (b) Hyperparasites

(c) Both (d) Auto parasite

49. Flare square symptom shown by

(a) American bollworm
(b) Spotted bollworm
(c) Pink bollworm
(d) All of the above

50. Cottony cushion scale is controlled by

(a) Vedalia beetle
(b) Mirid bug
(c) Hispid beetle
(d) Lady bird beetle

ANSWER KEY

1	2	3	4	5	6	7	8	9	10	11	12	13	14	15
(b)	(c)	(d)	(a)	(d)	(a)	(d)	(a)	(b)	(b)	(a)	(c)	(b)	(d)	(a)
16	17	18	19	20	21	22	23	24	25	26	27	28	29	30
(b)	(d)	(c)	(a)	(b)	(b)	(a)	(d)	(b)	(a)	(d)	(a)	(d)	(a)	(b)
31	32	33	34	35	36	37	38	39	40	41	42	43	44	45
(d)	(b)	(c)	(d)	(a)	(b)	(c)	(a)	(b)	(a)	(b)	(c)	(d)	(b)	(d)
46	47	48	49	50										
(a)	(b)	(c)	(b)	(a)										

Grand Test - 4

1. **Emperical formula of chlorophyll A is**
   (a) $C_{55}H_{70}O_6N_4Mg$ (b) $C_{55}H_{72}O_5N_4Mg$
   (c) $C_{40}H_{56}Mg$ (d) $C_{40}H_{56}O_2$

2. **Idea of gas exchange taking place in photosynthesis was given by**
   (a) Ingenhouz (b) Senebier
   (c) Sachs (d) Priestley

3. **More stomata on the lower surface of the leaves found in which type of stomata**
   (a) Apple type (b) Potato type
   (c) Oat type (d) Water lily type

4. **Imbibition in starchy seed is more than oily seed**
   (a) True (b) False

5. **Which of the following causes increase in female flowering**
   (a) GA (b) ABA
   (c) Ethrel (d) Maharashtra

6. **Precursor of gibbrellic acid is**
   (a) Mevalonic acid (b) Kaurene
   (c) Violaxanthin (d) Methionin

7. **Sand drawn disease of tobacco is due to**
   - (a) Mg deficiency
   - (b) Mg toxicity
   - (c) B deficiency
   - (d) B toxicity
8. **Which of the following is associated with reduction of plant growth**
   - (a) ABA
   - (b) Maharashtra
   - (c) Abscissic acid
   - (d) Cycocel
9. **Element which breaks water molecules during photosynthesis**
   - (a) Mg
   - (b) Mn
   - (c) Cu
   - (d) B
10. **Richmond Lang effect is related to**
    - (a) Auxin
    - (b) Gibberellin
    - (c) Cytokinin
    - (d) Ethylene
11. **For study of farming system, best PRA exercise is**
    - (a) Transect walk
    - (b) Resource mapping
    - (c) Chapati diagram
    - (d) Time line
12. ***Bhoodan Andolan* started in**
    - (a) 1951
    - (b) 1948
    - (c) 1958
    - (d) 1961
13. **Motto of *Pradhanmantri Krishi Sinchai Yojna***
    - (a) *Har fasal ko pani*
    - (b) *Khet khet ko pani*
    - (c) *Sookha na rahe koi khet*
    - (d) *Har khet ko pani*
14. **Objective of T and V system**
    - (a) To coordinate the improvement and extension activity
    - (b) To accelerate the improvement activities keeping and view the local need and situations
    - (c) To develop such training programs which can
    - (d) All of the above
15. **Communication is a**
    - (a) Method
    - (b) System
    - (c) Technique
    - (d) Process
16. **Which of the following works on single window delivery system**
    - (a) ATIC
    - (b) IRDP
    - (c) NAIP
    - (d) PMRY

17. MGNREGA was launched by

(a) Narendra Modi
(b) Atal Bihari Vajpeyi
(c) Manmohan Singh
(d) Indira Gandhi

18. Highest award of agricultural extension is

(a) Jawaharlal Nehru award
(b) Jagjeevan Ram innovative farmer award
(c) Krishi Karman Award
(d) Rafi Ahmad Kidwai award

19. *Pradhanmantri Jandhan Yojna* launched in

(a) 2013
(b) 2014
(c) 2015
(d) 2016

20. ATMA stands for

(a) Agriculture Technology Management Association
(b) Agriculture Technology Management Agency
(c) Agriculture Technology Mission Agency
(d) NOTA

21. Pusa arun is the seedless variety of

(a) Acid lime
(b) Kinnow
(c) Sweet lime
(d) Pumello

22. Marmalade is made from

(a) Apple
(b) Sweet orange
(c) Cashew
(d) Carrot

23. TSS in squash is (in percent)

(a) 40-50
(b) 60-65
(c) 55
(d) 75

24. Dwarf mutant of papaya is

(a) Pusa nanha
(b) Pusa giant
(c) Pusa dwarf
(d) None of the above of these

25. 1st Japanese style garden of India

(a) Nishant Bagh
(b) Buddha Jayanti Park
(c) Roshnara Park
(d) Rock Garden

26. **Pusa srijan is the rootstock for**
   (a) Guava (b) Ber
   (c) Citrus (d) Grape

27. **Lime induced chlorosis is caused by the deficiency of**
   (a) Fe (b) Zn
   (c) Mg (d) S

28. **Weeping jelly is due to excess of**
   (a) Sugar (b) Water
   (c) Acid (d) Pectin

29. **Degreening of citrus is done by**
   (a) NAA (b) IAA
   (c) GA (d) Ethylene

30. **Sinigrin,a toxic substance found in**
   (a) Solanum (b) Cole crops
   (c) Spinach (d) Lathyrus

31. **Nationalisation of RBI was done in which year**
   (a) 1935 (b) 1945
   (c) 1949 (d) 1951

32. **In production curve which stage is called double loss zone**
   (a) Stage 1 (b) Stage 2
   (c) Stage 3 (d) A and B

33. **Optimum theory of population was given by**
   (a) Malthus (b) Dalton
   (c) Canon (d) Sidgwick

34. **Utility which are created due to buying and selling**
   (a) Place utility (b) Time utility
   (c) Possession utility (d) Form utility

35. **Higher price is fixed so that rich buy it and then price is decreased is called**
   (a) Price penetration (b) Price skimming
   (c) Hedging (d) Price spread

36. Mahalwari system was introduced by

(a) William Bentick (b) Lord Curzon

(c) Thomas Munro (d) Dalhausie

37. VAT is an indirect tax

(a) True (b) False

38. The same property for which loan is taken,is offered for loan is called

(a) Mortgage loan (b) Hypothecation

(c) Key loan (d) Open loan

39. A statement of assets and liabilities of a farm at a point of time is called

(a) Income statement (b) Balance sheet

(c) Farm budgeting (d) Cash flow statement

40. Laissez-faire is a French word which means

(a) No role of government (b) Keeping others constant

(c) Is always greater than (d) None of the above of these

41. Local control is not applied for the design

(a) RBD (b) CRD

(c) LSD (d) SPD

42. For an asymmetrical data if mean=8 and median is 6 than value of mode is

(a) 2 (b) 8

(c) 0 (d) 12

43. 2nd central moment is equal to

(a) Mean (b) Skewness

(c) Kurtosis (d) SD

44. For symmetrical data value of kurtosis is

(a) <3 (b) 3

(c) >3 (d) None of the above

45. Which one of the following is not an absolute measure of dispersion

(a) Mean deviation (b) SD

(c) Variance (d) Coefficient of variation

46. ANOVA was defined by

(a) Pearson
(b) Fisher
(c) Secrist
(d) Bowly

47. Value of standard deviation is never be zero

(a) True
(b) False

48. Central tendency suitable for size of agricultural holdings

(a) Mean
(b) Median
(c) Mode
(d) All of the above

49. Chi square test was given by

(a) Fisher
(b) Karl Pearson
(c) Gosset
(d) De moivre

50. 2nd quartile is equal to

(a) Mid-point of the data
(b) Mid-point of the first half of the data
(c) Mid-point of the second half of the data
(d) None of the above

ANSWER KEY

1	2	3	4	5	6	7	8	9	10	11	12	13	14	15
(b)	(d)	(b)	(b)	(c)	(b)	(a)	(d)	(b)	(c)	(b)	(a)	(b)	(d)	(d)
16	17	18	19	20	21	22	23	24	25	26	27	28	29	30
(a)	(c)	(d)	(b)	(b)	(d)	(b)	(a)	(a)	(c)	(a)	(a)	(c)	(d)	(b)
31	32	33	34	35	36	37	38	39	40	41	42	43	44	45
(c)	(c)	(d)	(c)	(b)	(a)	(a)	(b)	(b)	(a)	(b)	(a)	(d)	(b)	(d)
46	47	48	49	50										
(b)	(a)	(c)	(b)	(a)										

Grand Test - 5

1. **Nursery area required to cultivate rice in 1 acre field**
   (a) 0.1 ha (b) 0.04ha
   (c) 0.4ha (d) 2.5acres

2. **The centre of origin of wheat according to vavilov**
   (a) Turkey (b) Mexico
   (c) New Guinea (d) Burma

3. **Agronomic measures are used to reduce erosion when slope is less than what percent**
   (a) 2 (b) 5
   (c) 10 (d) 20

4. **Word extension is derived from**
   (a) Latin (b) Greek
   (c) French (d) Arabic

5. **Training and visit system was first introduced in which state**
   (a) Rajasthan (b) Uttar Pradesh
   (c) Maharashtra (d) Karnataka

6. **Theory of profit was given by**
   (a) A walker (b) Malthus
   (c) Adam Smith (d) Lerner

7. Father of vital statistics

(a) Captain John Grant
(b) R.A. Fisher
(c) Karl Pearson
(d) Paul Leagans

8. Fine wool breed of sheep is native of

(a) Spain
(b) Pakistan
(c) Switzerland
(d) America

9. First Krishi Vigyan Kendra was under which university

(a) Indian Agricultural Research Institute
(b) Tamiln Nadu Agriculture University
(c) Punjab Agricultural University
(d) GB Pant Agriculture University

10. Which of the following is the correct statement

(a) Minikit program was developed in 1971
(b) NAFED was established in 1958
(c) Command area development authority was established in 1974
(d) All of the above

11. Which programme was developed on ICAR golden jubilee year

(a) Lab to land programme
(b) National rural employment programme
(c) Desert development programme
(d) 20-point programme

12. Nabard was established in the year 1982, while dwcra was established in 1980

(a) True
(b) False

13. First agriculture University in India was established in the year

(a) 1960
(b) 1974
(c) 1979
(d) 1975

14. National commission on agriculture was formed in the year

(a) 1970
(b) 1980
(c) 1940
(d) 1986

15. Jawaharlal Nehru award is given for best work in PhD in the field of agriculture and allied sciences

(a) True (b) False

16. NIOF is located at

(a) Uttar Pradesh (b) Haryana
(c) Uttarakhand (d) New Delhi

17. Red and purple colour of maize leaf is due to deficiency of

(a) Phosphorus (b) Potassium
(c) Nitrogen (d) Magnesium

18. Number of centre of origin is given by vavilov

(a) 8 (b) 6
(c) 10 (d) 9

19. How many time is ratoon crop recommended in sugarcane

(a) Once (b) 4 times
(c) 2-5 times (d) Twice

20. Agro climatic regional planning in India was initiated in which five year plan

(a) Seventh (b) Fifth
(c) Twelfth (d) Sixth

21. Country which has first commercialised hybrid rice technology

(a) India (b) China
(c) Denmark (d) USA

22. Camel crop is a common name of

(a) Bajra (b) Jowar
(c) Ragi (d) Maize

23. China ranks first in production and consumption of sugar in the world

(a) True (b) False

24. Grasss tetany in cattle is due to consumption of low

(a) Mg (b) Ca
(c) K (d) Na

25. Lathyrism is caused due to

(a) Benzo oxalin acetic acid (b) Benzo oxalin citric acid
(c) Benzene oxalic acetic chloride (d) Benzamine oxalic acetic acid

26. O horizon is absent in arable soils

(a) True (b) False

27. Which of the following is a passive factor

(a) Relief (b) Climate

(c) Biosphere (d) Age of land

28. Top most mineral horizon

(a) A horizon (b) O horizon

(c) AB horizon (d) B horizon

29. Zone of least weathering

(a) A horizon (b) B horizon

(c) O horizon (d) C horizon

30. Horizon found in forest areas

(a) O horizon (b) A horizon

(c) B horizon (d) C horizon

31. Qurtz constitutes how much percentage of Earth's crust

(a) 36 (b) 70

(c) 56 (d) 6

32. Edible part of pineapple

(a) Bracts (b) Receptacle

(c) Thalamus (d) Peduncle

33. Which of the following has fleshy receptacle as their edible part

(a) Fig (b) Bael

(c) Ber (d) Wood apple

34. Type of fruit in arecanut

(a) Drupe (b) Berry

(c) Nut (d) Sorosis

35. Schizocarpic fruit is found in

(a) Carrot (b) Beetroot

(c) Both (a) and (b) (d) Okra

36. Type of fruit in mulberry

(a) Sorosis (b) Syconus

(c) Nut (d) Berry

37. Which of the following is a tropical fruit

(a) Phalsa (b) Date palm

(c) Pine apple (d) Fig

38. Non climacteric fruits can ripe after removing from plants

(a) True (b) False

39. National horticultural board is located at

(a) Bangalore (b) Gurgaon

(c) New Delhi (d) Hyderabad

40. NRCSS is located at

(a) Ajmer (b) Jodhpur

(c) Jaipur (d) Nashik

41. Which of the following is a non-climacteric fruit

(a) Mango (b) Pineapple

(c) Pear (d) Sapota

42. Which of the following is a climacteric fruit

(a) Fig (b) Grape

(c) Litchi (d) Ber

43. Which of the following is a suitable intercrop for coconut

(a) Banana (b) Papaya

(c) Garlic (d) Turmeric

44. Complete metamorphosis occurs in

(a) Exopterygota (b) Endopterygota

(c) Apterygota (d) All pterygotes

45. Weeds are pests

(a) True (b) False

46. National centre for integrated pest management is located at

(a) New Delhi (b) Bangalore

(c) Uttar Pradesh (d) Hyderabad

47. Father of host plant resistance

(a) RH painter (b) Paul de bach

(c) NA cobb (d) Elcar

48. The pest density at which control measures should be applied to prevent an increasing pestpopulation from reaching the economic injury level is called as

(a) Economic injury level
(b) Damage threshold level
(c) Both (a) and (b)
(d) Action threshold level

49. Type of mouth part present in house fly

(a) Siphoning type
(b) Sponging type
(c) Piercing and sucking type
(d) Rasping and sucking type

50. Chewing and lapping type of mouthparts are present in

(a) Honey bee
(b) Housefly
(c) Thrips
(d) Grasshopper

ANSWER KEY

1	2	3	4	5	6	7	8	9	10	11	12	13	14	15
(b)	(a)	(a)	(a)	(a)	(a)	(a)	(a)	(b)	(d)	(a)	(b)	(a)	(a)	(a)
16	17	18	19	20	21	22	23	24	25	26	27	28	29	30
(a)	(a)	(a)	(a)	(a)	(b)	(b)	(b)	(a)	(a)	(a)	(a)	(a)	(d)	(a)
31	32	33	34	35	36	37	38	39	40	41	42	43	44	45
(a)	(a)	(a)	(b)	(a)	(a)	(c)	(b)	(b)	(a)	(b)	(a)	(a)	(b)	(a)
46	47	48	49	50										
(a)	(a)	(d)	(b)	(a)										

Grand Test - 6

1. **Dharampal singh was a famous**
   (a) Sorghum breeder (b) Cotton breeder
   (c) Oilseed breeder (d) Rice breeder

2. **Crop which has highest potassium uptake**
   (a) Potato (b) Rice
   (c) Tomato (d) Cotton

3. **IW/CPE ratio was given by**
   (a) Wilcox, 1974 (b) Parihar, 1964
   (c) Wilcox, 1964 (d) Parihar, 1974

4. **Lines joining points of constant moisture contents**
   (a) Isohyets (b) Isopluvial
   (c) Isostere (d) Isopleths

5. **Which of the following factor does not influence wind erosion**
   (a) Surface cover factor (b) Soil textural class factor
   (c) Field width factor (d) Slope gradient

6. **Spacing required for bunch type groundnut**
   (a) 30x10 cm (b) 40x10 cm
   (c) 30x20 cm (d) 30x15 cm

7. Hydraulic conductivity of soils varies directly with

(a) Non capillary porosity
(b) Total porosity
(c) Water holding capacity
(d) Capillary porosity

8. Cigarette is a non-virginia type tobacco.

(a) True
(b) False

9. Isoproturon belongs to which herbicide chemical group?

(a) Triazinones
(b) Phenyl ureas
(c) Dinitroanilines
(d) Diphenyl ethers

10. Mo per cent of ammonium molybdate is

(a) 51 per cent
(b) 41 per cent
(c) 33 per cent
(d) 46 per cent

11. Which element is the metal component of urease enzyme?

(a) Sodium
(b) Nickel
(c) Silicon
(d) Zinc

12. Application of clay to sandy soils is known as

(a) Rabbing
(b) Reclamation
(c) Marling
(d) Teasting

13. Soil order characterized by excess of weathering with dominant iron and aluminum oxide clays

(a) Alfisols
(b) Histosol
(c) Oxisols
(d) Mollisols

14. Dolomite is source of

(a) Ca and p
(b) Ca and B
(c) B and P
(d) Ca and Mg

15. Stem nodulating rhizobium sp. Is

(a) Azo-rhizobium
(b) Sino-rhizobium
(c) Bradyrhizobium
(d) All of the above

16. Which of the following is not a characteristic of black soil?

(a) Poor in N and P
(b) High in clay content
(c) Rich in organic matter
(d) Both (a) and (c)

17. Which of the following element is immobile in soil but mobile in plants?

(a) B
(b) Zn
(c) S
(d) P

18. Die-back of citrus is due to

(a) B (b) Fe
(c) Zn (d) Mn

19. 1st folic acid was found in spinach leaves

(a) True (b) False

20. Major process by which plants meet their P and K requirement.

(a) Mass flow (b) Diffusion
(c) Active transport (d) Root interception

21. When pH is low, solubility of Fe

(a) Increases (b) Decreases
(c) Remains constant (d) None of the above

22. Indicator plant for citrus greening

(a) Sweet orange (b) Mandrin orange
(c) Rangpur lime (d) Acid lime

23. Rosica is a mutant variety of

(a) Banana (b) Apple
(c) Grapes (d) Mango

24. Fasciation is a physiological disorder of

(a) Apple (b) Pineapple
(c) Citrus (d) Guava

25. Arka nishant is a variety of

(a) Carrot (b) Turnip
(c) Radish (d) Beetroot

26. Fruit juice per cent and TSS per cent in syrup are

(a) 20 per cent and 60 per cent (b) 25 per cent and 65 per cent
(c) 10 per cent and 60 per cent (d) 20 per cent and 65 per cent

27. Practice of collecting rubber latex is k/a

(a) Skiffing (b) Rabbing
(c) Withering (d) Tapping

28. Mangala is an improved variety of

(a) Arecanut (b) Cashewnut
(c) Pistachio nut (d) Walnut

29. Protandrous fruit is/are

(a) Aonla (b) Cashew nut

(c) Custard apple (d) Both (b) and (c)

30. San jose scale introduced in India from

(a) Australia (b) Italy

(c) Netherlands (d) England

31. Invitro synthesis of RNA

(a) Palade (b) Khorana

(c) Ochoa (d) Grifith

32. RNA polymerase found in

(a) Nucleolus (b) Nucleoplasm

(c) Nucleus (d) Mitochondria

33. Test cross ratio of complementary gene action

(a) 1:2:1 (b) 2:1:1

(c) 3:1 (d) 1:3

34. Linkage first observed in

(a) Sweet pea (b) Drosophila

(c) Maize (d) Arabidopsis

35. Maintenance of male sterility by sib-mating is adopted in

(a) GMS (b) CMS

(c) CGMS (d) None of the above

36. Purification of existing pureline variety can be done by

(a) Backcross method (b) Pedigree method

(c) Mass selection (d) Pureline selection

37. 1st gamma garden built in

(a) Calcutta (b) Delhi

(c) New zealand (d) Long island

38. Dichogamy is regarded as main cause for poor crop in

(a) Strawberry (b) Custard apple

(c) Plum (d) Loquat

39. A homozygous line developed by self-fertilization in a cp crop is termed as

(a) Pureline (b) Composite

(c) Inbred (d) Clone

40. In a gametophytic si, partial fertility can be attained by

(a) S1S1*S2S2 (b) S1S3*S2S3

(c) S1S3*S2S4 (d) S1S2*S3S4

41. C.O. for soft rot in potato

(a) *Macrophomina phaseoli* (b) *Rhizoctonia solani*

(c) *Erwinia carotovora* (d) *Sclerotium rolfsii*

42. Which of the following is/are maize disease?

(a) Turcicum leaf blight (b) Myrothecium leaf spot

(c) Dahiya disease (d) All of the above

43. Cotton leaf curl and chilli leaf curl are transmitted by ?

(a) Both by whitefly (b) Both by aphids

(c) Whitefly and aphid resp. (d) Whitefly and thrips resp.

44. Yellow dwarf disease of rice is caused by

(a) MLO (b) Virus

(c) Viroid (d) Fungi

45. Rusts can be treated by

(a) Carbendazim (b) Oxycarboxin

(c) Carboxin (d) Kitazin

46. Satellite virus discovered by

(a) Kassanis (b) Shepherd

(c) Doi *et al.* (d) T. O. Diener

47. Univoltine nematode

(a) Pratylenchus (b) Anguina

(c) Meloidogyne (d) Tylenchus

48. "Principle of nematology" book written by

(a) Chitwood and chitwood (b) Cobb

(c) Berkeley (d) Thorne

49. The seed act came into force in

(a) 1963
(b) 1966
(c) 1968
(d) 1969

50. Isolation distance for certified seed of okra

(a) 100
(b) 200
(c) 300
(d) 400

ANSWER KEY

1	2	3	4	5	6	7	8	9	10	11	12	13	14	15
(c)	(a)	(d)	(c)	(d)	(a)	(b)	(b)	(b)	(a)	(b)	(c)	(c)	(d)	(a)
16	17	18	19	20	21	22	23	24	25	26	27	28	29	30
(c)	(d)	(c)	(a)	(b)	(a)	(a)	(d)	(b)	(c)	(b)	(d)	(a)	(a)	(b)
31	32	33	34	35	36	37	38	39	40	41	42	43	44	45
(c)	(c)	(d)	(a)	(a)	(c)	(d)	(b)	(c)	(b)	(c)	(a)	(d)	(a)	(b)
46	47	48	49	50										
(a)	(b)	(d)	(d)	(b)										

Grand Test - 7

1. **Term guttation**
   (a) Burgerstein (b) Goldewski
   (c) Stephen Hales (d) Sachs

2. **Light saturation point is high in**
   (a) C3 (b) C4
   (c) CAM (d) None of the above

3. **The absorption of water by hydrophilic colloids is called**
   (a) Diffusion (b) Imbibition
   (c) Plasmolysis (d) Mass flow

4. **Which nutrient is related with water oxidizing enzyme complex?**
   (a) Mg (b) Zn
   (c) Cu (d) Mn

5. **Which gas is considered as antagonistic to Ethylene action?**
   (a) O_2 (b) N_2
   (c) CO_2 (d) CH_4

6. **IRDP was started in**
   (a) 1961 (b) 1964
   (c) 1968 (d) 1978

7. The rural systems research idea was motivated by MS Swaminathan in

(a) 1978 (b) 1968

(c) 1988 (d) 1998

8. Movie film recording is made by

(a) Mechanical (b) Optical

(c) Magnetic (d) All of the above

9. Indian Village Service by

(a) Mosher and Gupta (b) S K Dey

(c) Spencer Hatch (d) Wisher

10. Which of the following is not a stage of adoption

(a) Awareness (b) Evaluation

(c) Diffusion (d) Trial

11. Seminar is an example of

(a) Individual contact method (b) Group

(c) Mass/communit (d) None of the above

12. Protein per cent in dahi

(a) 3.2-3.4 per cent (b) 3.8-4.2 per cent

(c) 4.5-5 per cent (d) None of the above

13. Fat per cent in colustrum of sheep

(a) 4.5 per cent (b) 3.6 per cent

(c) 5.6 per cent (d) 7.4 per cent

14. Disease found in sheep during monsoon

(a) Foot rot (b) FMD

(c) Rinderpest (d) All of the above

15. Most dangerous disease of chicken?

(a) Chicken pox (b) White diarrhoea

(c) Mark's disease (d) Ranikhet

16. Splenic fever is common name for

(a) Anthrax (b) Rinderpest

(c) Fmd (d) Ranikhet

17. Lincoln is a breed of

(a) Australia (b) Switzerland

(c) England (d) France

18. Merino of Rajasthan

(a) Bikaneri (b) Chokla

(c) Marwari (d) Jaisalmeri

19. Throwing animals on ground is known as

(a) Culling (b) Docking

(c) Flushing (d) Casting

20. In the case of perfect negative correlation, the coefficient of correlation will be

(a) -1 (b) 0

(c) 1 (d) -1 to 1

21. Value of 0th central moment is

(a) 0 (b) 1

(c) Equal to mean (d) Varies

22. Term regression

(a) Fisher (b) Karl Pearson

(c) Francis Galton (d) Captain John Grant

23. Error degree of freedom for 6 replications in LSD.

(a) 36 (b) 25

(c) 20 (d) 12

24. Excretory product in red cotton bug

(a) Uric acid (b) Urea

(c) Ammonia (d) Allantoin

25. Amphipneustic condition found in

(a) Maggots (b) Chironomid larvae

(c) Coleopteran grubs (d) Lepidopteran larvae

26. Grandlure, a sex pheromone is produced by

(a) Pink bollworm (b) Cotton bollworm

(c) Cotton bollweevil (d) Cotton aphids

27. Geniculate antenna is present in

(a) Pulse bettle (b) Termites

(c) House fly (d) None of the above

28. Anti-juvenile hormone

(a) Ecdysone
(b) Buprofezin
(c) Percocenes
(d) Lufenuron

29. Degenerated ocelli of Cockroach is called

(a) Fenestra
(b) Acrotergite
(c) Glossa
(d) Both (a) and (b)

30. Trade name of chlorpyrifos

(a) Temik
(b) Dursban
(c) Baygon
(d) Dimecron

31. Skeletanization of tobacco leaves is caused by

(a) Aphid
(b) White fly
(c) Tobbaco caterpillar
(d) Tobacco stem gAll of the above

32. Processing leads to

(a) Form utility
(b) Space utility
(c) Time utility
(d) Possession utility

33. Rate of exchangeablility

(a) Price
(b) Wealth
(c) Value
(d) Possession

34. Curve for average variable cost is

(a) Inverse S shaped
(b) U shaped
(c) Parallel to X-axis
(d) Rectangular Hyperbola

35. Example for advalorem tax

(a) Land
(b) Excise duty
(c) Stamp paper
(d) Tractor

36. If rate of inflation varies from 3 per cent to 6 per cent,

(a) Creeping inflation
(b) Walking inflation
(c) Running inflation
(d) Galloping inflation

37. MRPS for complementary products

(a) <0
(b) >0
(c) 0
(d) None of the above

38. Profit gained by creating shortage in the market

(a) Hedging (b) Skimming

(c) Penetration (d) Speculation

39. First regulated market karanjia cotton market established in

(a) 1806 (b) 1826

(c) 1876 (d) 1886

40. With progress in technology, the PPC will shift

(a) Towards right (b) Toward left

(c) Remains same (d) None of the above

41. GDP =

(a) NDP- depreciation (b) NDP+ depriciation

(c) NNP-depriciation (d) NNP+depriciation

42. Term bacteria

(a) Lederberg (b) Leeuwenhoek

(c) Ehrenberg (d) Louis Pasteur

43. Polymorphism found in

(a) Mustard aphid (b) Rice gundhi bug

(c) Mango fruit flies (d) Sugarcane leaf hopper

44. Asymmetrical mouth parts are found in

(a) Termites (b) Weevils

(c) Mites (d) Thrips

45. The protective covering of sterile hyphae around an ascocarp is termed as

(a) Peridium (b) Epiderm

(c) Mycelium (d) Appendages

46. Ethylene during transportation of flowers causes

(a) Calyx splitting (b) Bent neck

(c) Drooping (d) Sleepiness

47. Servants of Indian Society, Pune was initiated by

(a) Albert Mayer (b) Mahatma Gandhi

(c) Col. Shitole (d) Gopal Krishna Gokhale

48. Communication process is stabilized by

(a) Empathy
(b) Credibility
(c) Feedback
(d) Channel

49. Father of agrometeorology

(a) Ramdas
(b) B Koppen
(c) Aristotle
(d) D N. Walia

50. The colour pigment involved in flowering in plant is

(a) Chlorophyll
(b) Cytochrome
(c) Phytochrome
(d) Anthocyanin

ANSWER KEY

1	2	3	4	5	6	7	8	9	10	11	12	13	14	15
(a)	(b)	(b)	(d)	(c)	(d)	(c)	(b)	(d)	(c)	(b)	(a)	(d)	(a)	(d)
16	17	18	19	20	21	22	23	24	25	26	27	28	29	30
(a)	(b)	(b)	(d)	(a)	(b)	(c)	(c)	(d)	(a)	(c)	(d)	(c)	(a)	(b)
31	32	33	34	35	36	37	38	39	40	41	42	43	44	45
(c)	(a)	(c)	(b)	(c)	(b)	(b)	(d)	(d)	(a)	(b)	(c)	(a)	(d)	(a)
46	47	48	49	50										
(d)	(d)	(c)	(d)	(c)										

Grand Test - 8

1. **Institution village linkage programme**
   (a) 1994 (b) 1995
   (c) 1997 (d) 1998
2. **National agriculture technology project**
   (a) 1998 (b) 1999
   (c) 1997 (d) 2004
3. **Kisan credit card scheme**
   (a) 1996 (b) 1997
   (c) 1998 (d) 1994
4. **National food for work program**
   (a) 2002 (b) 2003
   (c) 2004 (d) 2005
5. **Present president of icar**
   (a) Habibullah (b) Radha Singh Mohan
   (c) Piyush Goyal (d) Narendra Singh Tomar
6. **First Indian director of IARI**
   (a) B.P. Pal (b) G. Kallo
   (c) B. Vishvanath (d) Trilochan Mahapatra

7. Agricultural research service s started in

(a) 1972 (b) 1976

(c) 1973 (d) 1982

8. Lunishree var of

(a) Rice (b) Wheat

(c) Barley (d) Jowar

9. Concept of ideotype given by

(a) Glauber (b) Donald

(c) Stewart (d) Holliday

10. Jaya var of rice developed by

(a) Blackman (b) Kush

(c) Shastry (d) Michele

11. Removal of tail in sheep

(a) Culling (b) Casting

(c) Docking (d) Flushing

12. Blue tongue in cattle is

(a) Bacterial disease (b) Viral

(c) Fungal (d) Protozoa

13. Curve of normal distribution is

(a) Sigmoid (b) Bell shape

(c) L shape (d) Straight line

14. Chi square test given by

(a) Fisher (b) Gosset

(c) Karl pearson (d) Bernouli

15. Two-way classification and one way control

(a) CRD (b) RBD

(c) LSD (d) Both (a) and (b)

16. GST act came into effect from

(a) 1st July 2017 (b) 29th March 2017

(c) 29th April 2017 (d) 1st July 2018

17. Sales tax is example of

(a) Direct tax (b) Indirect tax

18. Demand for necessary goods is

(a) Elastic
(b) Inelastic

19. Demand for luxury goods is inelastic

(a) True
(b) False

20. Package program

(a) EDP
(b) IADP
(c) DPAP
(d) None of the above

21. Early majority category population

(a) 16 per cent
(b) 34 per cent
(c) 13.5 per cent
(d) 2.5 per cent

22. Father of sociology

(a) Brayne
(b) Spencer
(c) Albert
(d) Auguste

23. Krebs cycle occurs in

(a) Chloroplast
(b) Chromoplast
(c) Mitochondria
(d) Peroxisomes

24. Conversion of fat into sugars

(a) Lysosomes
(b) Glyoxisomes
(c) Peroxisomes
(d) None of the above

25. Guttation is caused due to root pressure

(a) True
(b) False

26. Deamination agents

(a) Acriflavin
(b) Bromouracil
(c) Ethium bromide
(d) EMS

27. Linkage of dominant allele with recessive allele is known as coupling

(a) True
(b) False

28. Tenalism is known as

(a) Translation
(b) Transcription
(c) Reverse transcription
(d) None of the above

29. DT46 is var of

(a) Wheat
(b) Triticale
(c) Maize
(d) Bajra

30. Brassica U triangle given by

(a) Pritam Kalia (b) Nagaharu
(c) Haberdt (d) Harrington

31. Translation is the process of

(a) RNA synthesis (b) Amino acid synthesis
(c) Protein synthesis (d) None of the above

32. Chloroplast discovered by

(a) Heackel (b) Schimper
(c) Flemming (d) Plowe

33. RNA discovered by

(a) Benda (b) Camilo
(c) Palade (d) De duve

34. Operon hypothesis

(a) Beadle (b) Jacob
(c) Lamarck (d) Darwin

35. Cancer of wheat

(a) Rust (b) Smut
(c) Karnal bunt (d) Powdery mildew

36. Wheat Flour gives fishy smell due to which disease

(a) Smut (b) Partial bunt
(c) Stem rust (d) None of the above

37. Cottony growth on leaf surface in barley is due to

(a) Covered smut (b) Loose smut
(c) Both (d) None of the above

38. Yellow slimy ooze on inflorescence of wheat is due to

(a) Tundu disease (b) Sehu
(c) Both (d) Hill bunt

39. According to IMMS total insect orders

(a) 28 (b) 29
(c) 27 (d) 26

40. Anamorphosis present in

(a) Diplura (b) Protura
(c) Collembola (d) Thysanura

41. Forewings modified as pseudo halters in

(a) Neuroptera (b) Diptera

(c) Strepsiptera (d) Coleoptera

42. Lady finger var of banana is

(a) Diploid (b) Triploids

(c) Tetraploid (d) None of the above

43. Honda is variety of

(a) Mandarin (b) Banana

(c) Papaya (d) Acid lime

44. Kokkan disease in banana is viral disease

(a) True (b) False

45. Mandarins are highly susceptible to waterlogging

(a) True (b) False

46. Total soil orders

(a) 10 (b) 11

(c) 12 (d) 13

47. Histosol is a

(a) Latin word (b) Greek

(c) French word (d) Arabic

48. Measuring solar radiation in narrow wave bands

(a) Pyrheliometer (b) Spectrophotometer

(c) Pyranometer (d) Heliometer

49. Dry adiabatic lapse rate is 6.5°C/km

(a) True (b) False

50. Actual seat of most photochemical reactions in air

(a) Troposphere (b) Stratosphere

(c) Ionosphere (d) Thermosphere

ANSWER KEY

1	2	3	4	5	6	7	8	9	10	11	12	13	14	15
(b)	(a)	(c)	(c)	(d)	(c)	(c)	(a)	(b)	(c)	(c)	(b)	(b)	(c)	(b)
16	17	18	19	20	21	22	23	24	25	26	27	28	29	30
(a)	(b)	(b)	(b)	(b)	(b)	(d)	(c)	(b)	(a)	(c)	(b)	(b)	(b)	(b)
31	32	33	34	35	36	37	38	39	40	41	42	43	44	45
(c)	(b)	(c)	(b)	(c)	(b)	(b)	(c)	(b)	(b)	(c)	(a)	(b)	(a)	(a)
46	47	48	49	50										
(c)	(b)	(b)	(b)	(b)										

Grand Test - 9

1. **Stale seed bed technique of weed control is a ______ method.**
   (a) Cultural
   (b) Mechanical
   (c) Chemical
   (d) Biological

2. **PF refers to______**
   (a) Logarithm of H ion concentration
   (b) Logarithm of free flow water
   (c) Logarithm of soil moisture tension
   (d) Logarithm of salt concentration

3. **White rust of mustard is caused by**
   (a) *Puccinia graminis*
   (b) *Puccinia recondita*
   (c) *Puccinia arachidis*
   (d) *Albugo candida*

4. **Total root parasite of tobacco is**
   (a) Cuscuta
   (b) Orobanche
   (c) Striga
   (d) Loranthus

5. **The whip smut of sugarcane is caused by**
   (a) *Ustilago maydis*
   (b) *Ustilago scitaminea*
   (c) *Puccinia*
   (d) None of the above

6. **The symptoms of red rot of sugarcane mostly found on**
(a) Stem (b) Leaf
(c) Root (d) Arrow

7. **Frog-eye leaf spot disease is present in**
(a) Tobacco (b) Cotton
(c) Potato (d) None of the above of above

8. **The first recognized virus disease of plants in world is**
(a) Tobacco mosaic (b) Yellow vein mosaic
(c) Cucumber mosaic (d) None of the above of above

9. **Solar heat treatment is used to control**
(a) Covered smut of barley (b) Smut of bajra
(c) Loose smut of wheat (d) Smut of sorghum

10. **Phyllody is caused by**
(a) Bacteria (b) Virus
(c) Fungi (d) Phytoplasma

11. **Which of the following rust is not caused by fungi ?**
(a) Black rust (b) Brown rust
(c) Yellow rust (d) Red rust

12. **The alternate host for pearl millet rust is**
(a) Brinjal (b) Tomato
(c) Mustard (d) Okra

13. **Fungi are**
(a) Autotrophic (b) Heterotrophic
(c) Phototrophic (d) None of the above

14. **Ascus normally contains definite number of ascospores which is**
(a) 2 (b) 4
(c) 6 (d) 8

15. **The gladiator belongs to the insect order**
(a) Phasmida (b) Orthoptera
(c) Grylloblattoidea (d) Mantophasmatodea

16. **The differentiation of the body into distinct functional regions is called**
(a) Morphogenesis (b) Oogenesis
(c) Tagmosis (d) Blastogenesis

17. In 2022, world pest day celebrated on

(a) 6 june
(b) 7 june
(c) 6 may
(d) 7 may

18. In which type of head the mouth parts are in continuous series with thelegs?

(a) Hypognathous
(b) Prognathous
(c) Ophisthognathous
(d) Paurometabolous

19. The Johnston's organ is present on

(a) Scape
(b) Pedicel
(c) Coxa
(d) Trochanter

20. In silk worm, the antennae are

(a) Clavate
(b) Capitate
(c) Pectinate
(d) Bipectinate

21. Foregut is lined by

(a) Intima
(b) Peritrophic membrane
(c) Enteric epithelium
(d) Taenidia

22. Salivary glands are also called

(a) Mandibular glands
(b) Maxillary glands
(c) Labial glands
(d) Pharyngeal glands

23. The neurons which carry impulses from the central nervous system are called

(a) Sensory
(b) Afferent
(c) Motor
(d) Interneurons

24. The sac for storage of sperms in male insect is called

(a) Spermatheca
(b) Seminal vesicle
(c) Median oviduct
(d) Genital chamber

25. The amount of uric acid in excreta of insects is about

(a) 10 per cent
(b) 25 per cent
(c) 40 per cent
(d) 85 per cent

26. Juvenile hormone is secreted by

(a) Neurosecretary cells
(b) Prothoracic glands
(c) Corpora allata
(d) Corpora cardiaca

27. What is the award name which has given to the states for good agricultural production by prime minister Narendra MModi?

(a) Krishi kirti (b) Krishi karman

(c) Krishi samman (d) Krishi samridhi

28. Which prize is considered as nobel prize of agriculture?

(a) World food prize (b) Rafi Ahmed Kidwai award

(c) Wolf prize (d) None of the above

29. "The limiting velocity of a particle falling in fluid medium will be proportional to the squareof its diameter." This low is given by:

(a) Beer (b) Jackson

(c) Stoke (d) Lambert

30. This enzyme is involved in the Biological nitrogen fixation

(a) Rhizobium (b) Azotobacter

(c) Nitrogenase (d) Nitrosomonas

31. "Enrichment of water bodies by nutrients and consequent bloom of algal and other aquatic plants" This is the process known as:

(a) Illuviation (b) Elluviation

(c) Eutrophication (d) Neutralization

32. A branch of soil science which deals with the soil from the crop production point of view

(a) Pedology (b) Soil geology

(c) Ecology (d) Edaphology

33. Particle density is also known as:

(a) True density (b) Bulk density

(c) Apparent density (d) Specific gravity

34. Total porosity is more in

(a) Sandy soil (b) Clayey soil

(c) Silty soil (d) Loamy sand

35. The best agricultural soil structure is

(a) Crumb (b) Platy

(c) Columnar (d) Bloky

36. If igneous rock contains more than 65 per cent silica it is called as

(a) Acidic (b) Intermediate

(c) Basic (d) Ultrabasic

37. The soil having more than 30 per cent organic matter is placed in

(a) Oxisols (b) Aridisols

(c) Histosols (d) None of the above

38. Infiltration rate is relatively higher in

(a) Sandy soil (b) Clay loam soil

(c) Silty soil (d) Clayey soil

39. Widely accepted fixed value of particle density is

(a) 1.5 g/cc (b) 2.65 g/cc

(c) 2.29 g/cc (d) 1.65 g/cc

40. Which one is considered as organic fertilizer?

(a) FYM (b) Urea

(c) Ammonium sulphate (d) DAP

41. kg/ha = ppm x ______

(a) 2.24 (b) 4.48

(c) 1.12 (d) 1.24

42. The parent materials of sand size transported by the action of wind is called

(a) Eolian (b) Moraine

(c) Loess (d) Alluvium

43. Which of the followings is a measure of central value?

(a) Median (b) Standard deviation

(c) Mean deviation (d) Quartile deviation

44. Which of the followings represents median?

(a) First quartile (b) Fiftieth percentile

(c) Sixth decile (d) None of the above

45. If the minimum value in a set is 9 and its range is 57, the maximum value of the set is

(a) 33 (b) 66

(c) 48 (d) None of the above

46. The average of the sum of squares of the deviations about mean is called:

(a) Variance (b) Absolute deviation

(c) Standard deviation (d) Mean deviation

47. Which of the following is unit less measure of dispersion ?

(a) Standard deviation (b) Mean deviation

(c) Coefficient of Variation (d) Range

48. The hypothesis under test is:

(a) Simple hypothesis (b) Alternative hypothesis

(c) Null hypothesis (d) None of the above

49. Genotypes of zygotes of cross- pollinated cross are:

(a) Homozygous (b) Heterozygous

(c) Homogenous (d) Hetrogenous

50. Single plant progeny of self-pollinated crop is :

(a) Homozygous (b) Heterozygous

(c) Homogenous (d) Heterogenous

ANSWER KEY

1	2	3	4	5	6	7	8	9	10	11	12	13	14	15
(a)	(c)	(d)	(b)	(b)	(a)	(a)	(a)	(c)	(d)	(d)	(a)	(b)	(d)	(d)
16	17	18	19	20	21	22	23	24	25	26	27	28	29	30
(c)	(a)	(a)	(b)	(d)	(a)	(c)	(c)	(b)	(d)	(c)	(b)	(c)	(c)	(c)
31	32	33	34	35	36	37	38	39	40	41	42	43	44	45
(c)	(d)	(a)	(b)	(a)	(a)	(c)	(a)	(b)	(b)	(a)	(a)	(a)	(b)	(b)
46	47	48	49	50										
(a)	(c)	(c)	(b)	(c)										

Grand Test 10

1. **Father of zero tillage**
   (a) G.B. Triplets
   (b) Peter de
   (c) Liebig
   (d) Jethrotal
2. **Highest pesticide consumption state in India is**
   (a) Tamil Nadu
   (b) Punjab
   (c) Andhra Pradesh
   (d) Maharashtra
3. **India rank in coffee production**
   (a) 6
   (b) 3
   (c) 2
   (d) 4
4. **Highest pulse producing state is**
   (a) Tamil Nadu
   (b) Maharashtra
   (c) Madhya Pradesh
   (d) West Bengal
5. **SRI seed rate per acre**
   (a) 4-5
   (b) 3-4
   (c) 6-7
   (d) 2-3
6. **Forest area of India is**
   (a) 21.71 per cent
   (b) 20 per cent
   (c) 20.4 per cent
   (d) None of the above

7. CICR located in

(a) Nagpur (b) Coimbatore
(c) Jabalpur (d) None of the above

8. Highest fertilizer consumption state in India is

(a) Tamil Nadu (b) Andhra Pradesh
(c) Pondicherry (d) Punjab

9. National research centre for weed science located in

(a) Jodhpur (b) Junagarh
(c) Jabalpur (d) Jaipur

10. Chemically vitamin B2 is

(a) Riboflavin (b) Ascorbic acid
(c) Thiamine (d) Biotin

11. Verges kurien got world food prize in

(a) 1988 (b) 1986
(c) 1985 (d) 1989

12. Root formation

(a) GA3 (b) IBA
(c) IAA (d) ABA

13. National horticulture mission initiated in

(a) July 2006 (b) May 2005
(c) June 2005 (d) May 2006

14. PPV and FR established in

(a) Nov 2005 (b) June 2005
(c) July 2006 (d) May 2006

15. Rank order of countries in transgenic crop production

(a) USA>BRAZIL>INDIA>CUBA
(b) USA>BRAZIL>ARGENTINA>CANADA
(c) USA>BRAZIL>CUBA>CANADA
(d) USA>BRAZIL>INDIA>CANADA

16. 1st national research institute is for ______ located in ______

(a) Sorghum, Tamil Nadu (b) Groundnut,Junagarh
(c) Wheat, Ooty (d) Rice, Cuttack

17. Prevents premature fruit dropping

(a) GA3
(b) IAA
(c) ABA
(d) IBA

18. IIVR located in

(a) Modipuram
(b) Hyderabad
(c) Varanasi
(d) Jaipur

19. State having maxim area under irrigation is

(a) Punjab
(b) Tamil Nadu
(c) Maharashtra
(d) Karnataka

20. Father of golden revolution

(a) K.L. Chadda
(b) M.S. Swaminathan
(c) R.S. Paroda
(d) G.S. Khush

21. Father of white revolution

(a) Verges Kurien
(b) M.S. Swaminathan
(c) K.L. Chadda
(d) G.S. Khush

22. Chemically vitamin C is

(a) Tocopherol
(b) Biotin
(c) Thiamine
(d) Ascorbic acid

23. Order of transgenic crop production in the world

(a) Maize>soyabean>canola>cotton
(b) Soyabean>maize>cotton>canola
(c) Maize>soyabean>cotton>canola
(d) Maize>soyabean>cotton>brinjal

24. Sprout stop is the trade name of

(a) GA3
(b) IAA
(c) Mallic hydrazide
(d) IBA

25. NBPGR located in and founded in

(a) Hyderabad, 1976
(b) Uttar Pradesh, 1975
(c) Rajasthan, 1986
(d) Delhi, 1976

26. The relationship between plant population and yield is given by

(a) Response curve
(b) Biological growth
(c) Physiological growth
(d) None of the above

27. Hormone as well as herbicide

(a) GA3 (b) ABA

(c) 2,4-D (d) IAA

28. First all India coordinated research project on which crop (AICRP) in 1957

(a) Wheat (b) Rice

(c) Maize (d) Sorghum

29. 1st rice hybrid of India is

(a) MGR-1 (b) CORH-1

(c) Both (a) and (c) (d) PRH-1

30. Beta carotene rich rice variety is

(a) Golden rice (b) Biofortified rice

(c) Super rice (d) Red rice

31. Total number of KVKs at present

(a) 670 (b) 716

(c) 711 (d) 722

32. Occurence of Irish famine in

(a) 1835 (b) 1845

(c) 1943 (d) 1855

33. India rank in wheat and rice production

(a) 4 (b) 3

(c) 5 (d) 2

34. NCIPM stands for

(a) National centre for integrated pest management

(b) National centre for integrated post-harvest management

(c) National centre for integrated post technology management

(d) National centre for integrated crop pest management

35. Most pollutant in water is

(a) Arsenic (b) Mercury

(c) Cadmium (d) Selenium

36. Super rice developed by

(a) G.S. Khush (b) R.S. Paroda

(c) M.S. Swaminathan (d) K.L. Chadda

37. M.S.Swaminathan got world food prize in

(a) 1987
(b) 1983
(c) 1986
(d) 1988

38. Chemically vitamin B12 is

(a) Biotin
(b) Ascorbic acid
(c) Thiamine
(d) Cobalamine

39. Induction of dormancy

(a) IAA
(b) GA3
(c) IBA
(d) ABA

40. ICAR day is celebrated in

(a) July 16
(b) June 16
(c) May 16
(d) None of the above

41. Crop insurance scheme initiated in

(a) 1996
(b) 1985
(c) 1981
(d) 1980

42. Bollgaurd 2 gene is

(a) Cry 2 Aa and cry 2 Ab
(b) Cry 1Ac and cry 2 Ab
(c) Cry 2 Ab
(d) None of the above

43. NABARD was found in

(a) 1980
(b) 1981
(c) 1990
(d) 1982

44. International year of 2020 for

(a) Plant health
(b) Pulses
(c) Indigeneous year
(d) Oilseeds

45. Geographical area of India

(a) 310.74 mha
(b) 328.74 mha
(c) 330.24 mha
(d) 325 mha

46. Bollgaurd 1 gene is

(a) Cry 2 Ab
(b) Cry 1Ac
(c) Cry 2 Aa
(d) None of the above

47. National rice research institute located at

(a) Jabalpur (b) Hyderabad

(c) Cuttack (d) Punjab

48. Bhopal gas tragedy happened in

(a) 1984 (b) 1983

(c) 1884 (d) 1920

49. Enzyme is given by

(a) Went (b) W. Kuhne

(c) De morris (d) Funk

50. State with maximum area under drip irrigation is

(a) Maharashtra (b) Tamil Nadu

(c) West Bengal (d) Punjab

ANSWER KEY

1	2	3	4	5	6	7	8	9	10	11	12	13	14	15
(a)	(c)	(a)	(c)	(b)	(a)	(a)	(c)	(c)	(a)	(d)	(b)	(b)	(a)	(b)
16	17	18	19	20	21	22	23	24	25	26	27	28	29	30
(b)	(b)	(c)	(a)	(a)	(a)	(d)	(b)	(c)	(d)	(a)	(c)	(c)	(c)	(a)
31	32	33	34	35	36	37	38	39	40	41	42	43	44	45
(d)	(b)	(d)	(a)	(a)	(a)	(a)	(d)	(d)	(a)	(b)	(b)	(d)	(a)	(b)
46	47	48	49	50										
(b)	(c)	(a)	(b)	(a)										